BUILDING A READY-TO-RUN MODEL RAILROAD

— A quick and easy layout from off-the-shelf components —

Jeff Wilson

Alfred Noble Library
32901 Plymouth Road
Livonia, MI 48150-1793
(734) 421-6600

625.1
W

SEP 04 2009

Kalmbach Books
21027 Crossroads Circle
Waukesha, Wisconsin 53186
www.kalmbach.com/books

© 2008 by Kalmbach Books
All rights reserved. Except for brief excerpts for review, this book may not be reproduced in part or in whole by electronic means or otherwise without written permission of the publisher.

Published in 2008
12 11 10 09 08   1 2 3 4 5

Manufactured in the United States of America

ISBN: 978-0-89024-740-2

3 9082 11320 2702

Publisher's Cataloging-In-Publication Data
(Prepared by The Donohue Group, Inc.)

Wilson, Jeff, 1964-
Building a ready-to-run model railroad : a quick and easy layout from off-the-shelf components / Jeff Wilson.
p. : ill. ; cm.
ISBN: 978-0-89024-740-2
1. Railroads--Models. 2. Railroads--Models--Design and construction. 3. Models and modelmaking. I. Title.
TF197 .W5476 2008
625.1/9

Contents

Alfred Noble Library
32901 Plymouth Road
Livonia, MI 48150-1793
(734) 421-6600

The expanding variety of ready-to-use products makes it easier now than ever to get a realistic layout up and running in a short time. The locomotive is an Alco RS-27 from Proto 2000, the refrigerator car is from InterMountain, and the automobile is from Mini Metals.

Introduction

For many years, the term *ready-to-run* usually had negative connotations. Other than expensive imported brass locomotives and freight cars, most ready-to-use models fell into the toy category. These locomotives and cars had unrealistic body styles, heavy and out-of-scale details, and poorly operating horn-hook couplers and plastic wheelsets. To get good-quality rolling stock models, you had to buy and assemble kits, whether they were plastic, wood, resin, or another material, and to get good locomotives you had to be ready to add loads of aftermarket detail items.

This began to change in the 1990s when Kadee released its then-revolutionary ready-to-run model of a 40-foot Pullman-Standard PS-1 boxcar. The model featured many separately added details, realistic paint schemes with minute, legible lettering, metal wheels, and Kadee's own knuckle couplers, which had been the de facto standard among serious modelers for years.

That boxcar's success soon led other manufacturers to produce highly detailed models. Ready-to-run locomotives now feature lots of separate details, many based on specific railroad prototype variations. Rolling stock is also available with prototype-specific details featuring accurate, sharply rendered paint and lettering.

Along with the trains themselves, modelers now have access to high-quality assembled structures, vehicles, trees, and other scenic items.

Modeling realistic track once required laying track on roadbed and adding separate loose ballast. Several lines of track now include the roadbed with ballast profile and texture molded in place. Although most serious modelers still lay track the traditional way, the new all-in-one track and roadbed makes it much easier to get trains up and running.

Control systems have also improved greatly in recent years. The increasing popularity of Digital Command Control (DCC) has resulted in simplified, easy-to-operate systems, and many locomotives come from the factory equipped with DCC decoders and sound systems.

This book describe how to build a layout from start to finish using a variety of ready-to-use products. It's built in HO scale, which has the widest variety of available products. You can use similar benchwork and scenery methods for N and other scales, but the specific products available to you will

The completed layout is attractive and fits well in a rec room or other family area.

be different. Each chapter is designed to explain how to complete a specific step, from benchwork and tracklaying through structures and scenery to final detailing.

The layout shown in this book is set in the Midwest during the 1960s, but you certainly don't have to build the exact layout – in fact, I encourage you to alter it to fit your own interests. Set your railroad in a different locale and era, use different structures, try other types of scenery, and select the locomotives and rolling stock that appeal to you.

Use the book as a guide to building your model railroad. After completing all the outlined steps, not only will you have a completed layout, but you will gain an understanding of your modeling skills and interests. Where possible, I have included several options for a project, with varying degrees of involvement, so you can go beyond the basics if you like. Don't feel like you have to complete every step of every project. For example, if you find that you don't enjoy making custom signs for a structure, you can stick with the ones that come in the kit.

There are many other methods and techniques out there for completing the projects, so feel free to experiment on your own. In several places, I suggest other reference books that will help you discover additional tips and techniques.

Although I've used as many ready-to-run products as possible, model railroading still requires some basic construction, assembly, and detail work. The benchwork, or train table, still needs to be built and formed and scenery materials need to be applied. Don't be afraid – although these steps take a bit of work, today, it's much easier and faster to get a layout built and trains running than it was even a few years ago.

A word about products: Some items, such as green ground foam, are almost always available, but structures, locomotives, and rolling stock are often limited in production. There's no way to guarantee that all of the products I used will available as you read this, so don't be afraid to substitute similar items. But if a model has been produced, it will still be available somewhere. With a little research you'll be able to find virtually any product that has ever been made, whether in a hobby shop or on-line auction site such as eBay. Product resources are listed at the back of the book.

Now, let's move on to the first step and plan and build the layout's train table.

1-1

CHAPTER ONE

Track planning and benchwork

The benchwork for the layout consists of foam sheets on a wood frame placed on bookcases.

It's tempting to jump in and begin building a layout right away, when your excitement level is high. However, it's wise to take some time in planning, so the result will be a layout that you'll be happy with long after it's complete. Jumping in too quickly can lead to poor construction quality or a bad track plan, which can result in a layout never getting finished or simply sitting and gathering dust. Solid benchwork is also important, and today, it's easy to do with a minimum of tools, 1-1.

Track planning

The first step is to develop a track plan that fits your available space. You'll need to consider several factors, including the size of your room or available area, the dimensions of the layout itself, the specific railroad and era you want to model, the types of trains you want to run, and the style of terrain your layout will feature.

It's usually best to start with a published track plan, especially if you've never designed or built a layout before. You can then modify the track arrangement, structures, and scenic treatment to suit your needs.

The 4 x 8-foot layout has been a standard size for starter layouts since the early days of the hobby. It's a natural choice, since it's a common size for plywood and foam insulation board. Also, many published track plans have been designed for this space.

At first, a 4 x 8-foot table seems like a fairly compact layout size, but keep in mind that the four-foot width requires leaving both sides open to provide access, so you can reach across the table. Even when allowing two feet of room on each side of the layout – the bare minimum for comfortably moving around it – a 4 x 8 layout easily requires a space of at least 8 x 12 feet, and preferably more. And suddenly, it doesn't look quite so compact.

For this reason, layouts that can be placed along a wall, or walls, are popular choices. While allowing more open room around them, they also can provide more actual layout space. This was an important consideration for me, as I wanted to place the layout in a basement rec room, **1-2**.

I based the design of this layout on the Marquette & Independence, an HO model railroad (described in the December 1975, February 1976, and April 1976 issues of *Model Railroader* magazine). The layout is formed by diagonally cutting a 4 x 8-foot sheet of foam in two, **1-3**. Flopping one end and connecting the two pieces with a short section creates a dogbone shape with a loop at each end.

The layout was designed to fit along two walls of a basement rec room.

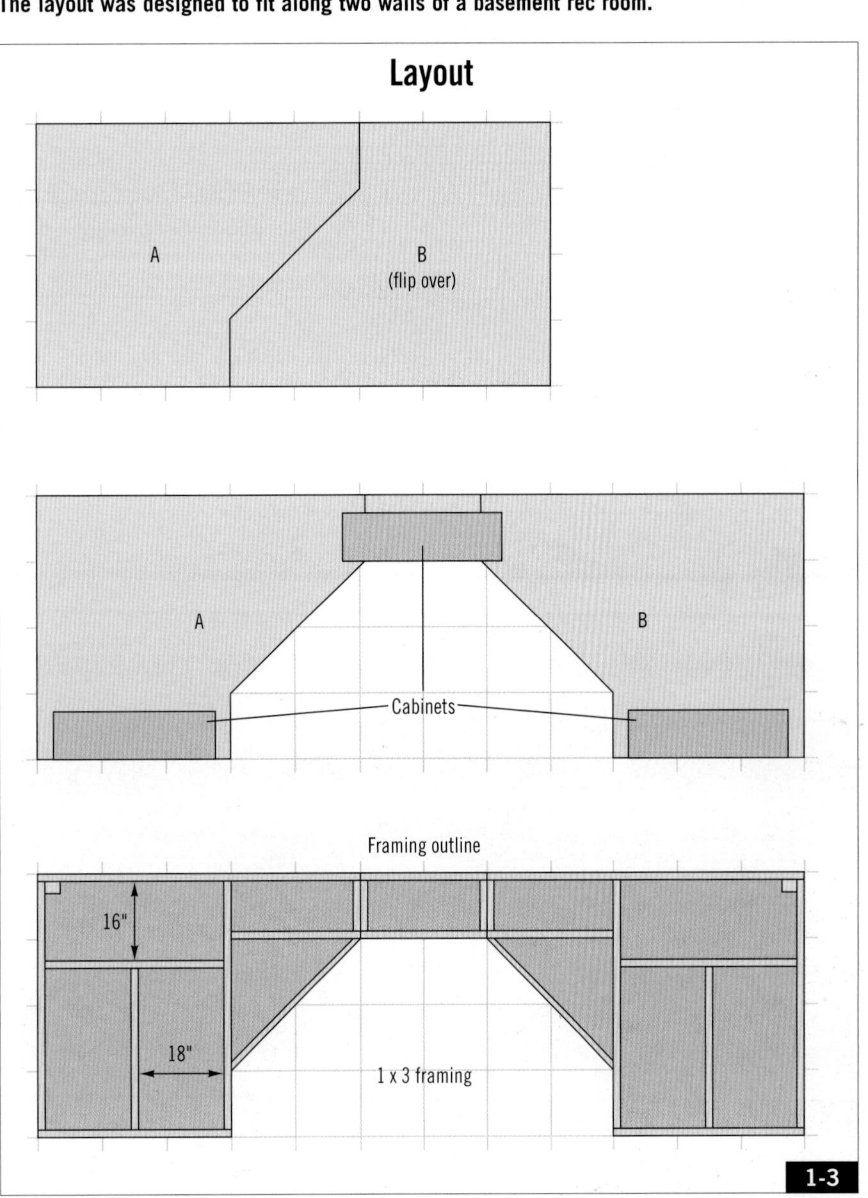

The layout design is simple, but it provides a relatively long mainline run for the extra area that the layout occupies.

I like this plan for several reasons. The design itself – with the operator's bay inside the layout's outline – allows the long, flat side (and both ends, if desired) to be placed against a wall. This greatly increases the useable space in a room, even though the layout has only slightly more surface area than a 4 x 8-foot table. The two ends of the layout are separate and distinct, allowing different scenic treatments in each, which can be difficult to accomplish on a 4 x 8 table layout. The resulting mainline run is also longer than possible on a conventional table, **1-4**.

The layout can be expanded, either by extending the connector piece between the ends (which can be as long as desired) or by adding a branch off one or both ends. I decided to use a 1 x 2-foot connector piece.

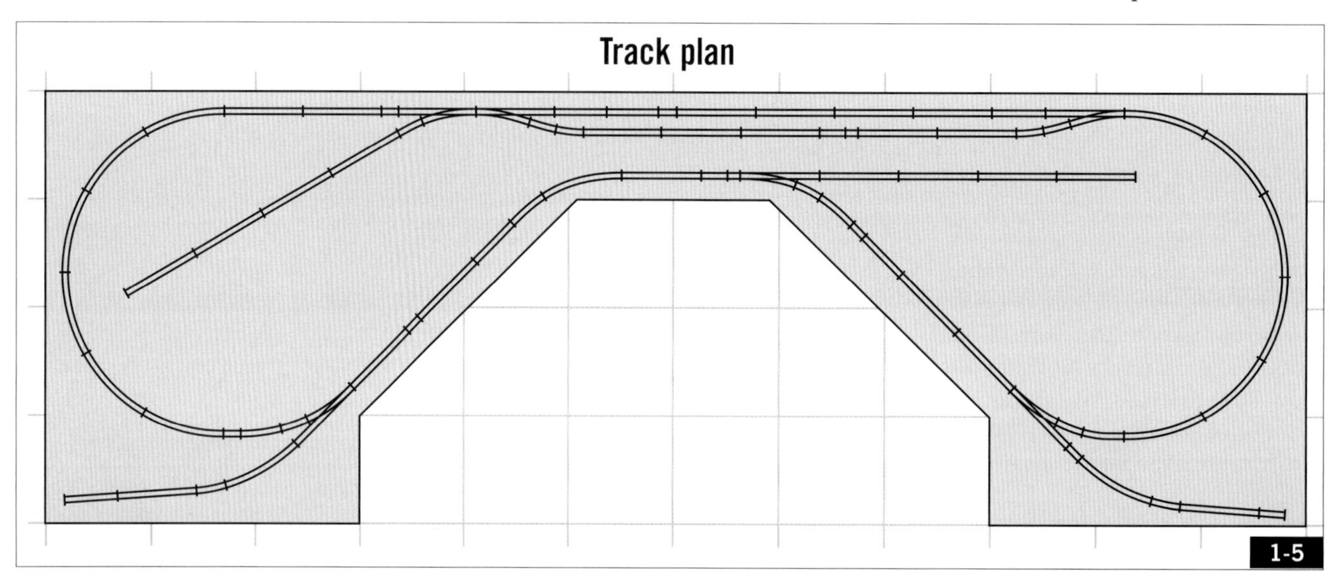

Mark the outline of the layout on the floor with masking tape and experiment with various track arrangements.

A bookcase makes a solid foundation for the layout table and offers built-in storage space.

The track plan is simple: essentially a large oval with one double-ended siding and several stub-ended spurs, **1-5**. I think it's beneficial for beginning and intermediate layouts to have a continuous run. The siding serves as a runaround track, allowing operators to easily switch cars in either direction at the various spurs, or it can be used to hold a second train.

Although larger than a conventional table layout, this one isn't incredibly complex, and it allows a beginning or intermediate modeler the chance to complete a layout in a reasonable time.

Once I had the size and basic plan figured out, I outlined the layout on the floor with masking tape, **1-6**. Doing this allows you to see exactly how the layout will fit into your space, and it's much easier to make adjustments now rather than when you're cutting benchwork components.

With the layout outline marked, get out your track and make sure your plan fits the layout space the way you want it to. You'll have time to adjust details of the track plan later, but now you should make sure that the mainline curves fit within the limits of the layout. When drawing a track plan on paper, it's easy to overestimate how much track fits into a given space – now's the time to double-check.

Benchwork

Benchwork has one basic function: to keep your layout from falling to the floor. It should be strong enough to absorb a bump or two without moving, sturdy enough to not sag over time, provide a solid surface for track and scenery, and – if possible – allow for underneath storage. Some modelers go overboard and build benchwork that's even stronger than a house wall, but this is generally a waste of time and lumber.

I wanted a layout that could be placed in a basement rec room, so the final appearance of the layout was important. I also wanted to use materials that are easy to work with, readily available, and simple to cut – features that will appeal to apartment dwellers and those who don't have a shop full of power tools.

1-8 The layout's design allows an an operator to sit comfortably at each end of the layout while having plenty of space available for storing larger items under the layout table.

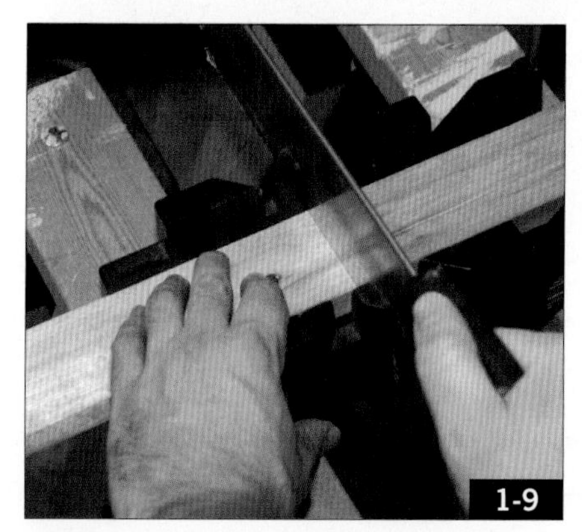

1-9 A miter saw ensures straight cuts and a square frame.

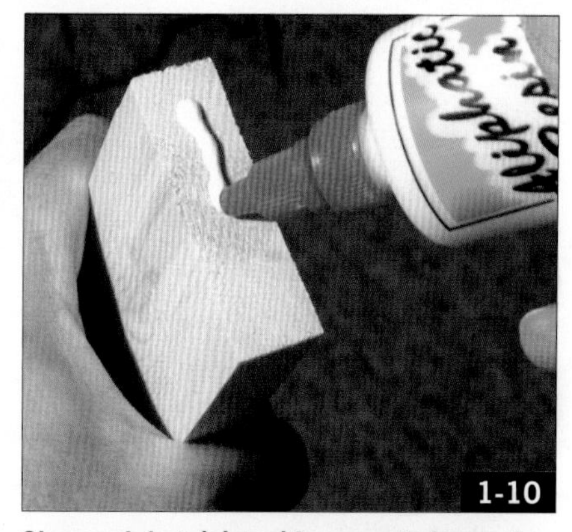

1-10 Glue each butt joint with carpenter's wood glue for a solid, finished frame.

1-11 Drill pilot holes in the ends of the 1 x 3s and secure the joint with wood or drywall screws.

Add an adjustable foot and T-nut to the bottom of each leg.

Clamp the leg in place and add two wood screws though the frame to secure the leg.

The framework rests on the bookcases, with the leg supporting the back corner. Cardboard protects the bookcase tops.

To that end, I built the layout atop three commercial bookcases from Office Depot. These bookcase kits (no. 636-140) are easy to build and relatively inexpensive. They provide a sturdy foundation for the layout and give it a classy appearance. You can also store trains and books or display memorabilia on them, **1-7**.

I used extruded foam insulation board for the layout surface. Foam is lightweight, sturdy, readily available, and can be cut and shaped with hand tools.

Start the benchwork by assembling the bookcases. Most bookcase kits are easily assembled with hand tools. I only needed a screwdriver and a hammer (for tacking the back panels in place). The brand of bookcase kit isn't important – the key is to find three identical models at the height you prefer. Mine measure 44⅛" tall, 27¾" wide, and 11½" deep. This means that the finished layout will be about 50" off the floor, which for me is a comfortable height to operate from while sitting on a stool. If this seems tall for you, just use shorter bookcases, **1-8**.

1-15

Fasten the framework sections together with two bolts, making sure the tops of abutting sections are aligned.

Framing

Resting on the bookcases, a skeleton frame, made of 1 x 3 boards, supports the foam table. Although foam sheets (especially the two-inch thick ones) are quite rigid, they can bow a bit if they're not framed. Attaching foam to a light frame keeps it flat, adds strength, and makes it easier to join sections together. The 1 x 3s are a bit of overkill since 1 x 2s would work just fine, but on the day I went to the lumberyard, the 1 x 3s were straighter and of better quality, so I opted for the larger lumber. Be sure the boards you select are straight and without any twists, or it will be impossible to build square, level frames.

Cut the 1 x 3s to size as shown in **1-3**. You can use a power miter saw if you have one or a miter box and hand saw, **1-9**. Many lumberyards and home centers will also make cuts for you or provide a saw for your use.

Glue each joint using carpenter's wood glue, **1-10**. Then, secure each joint with a pair of wood or drywall screws, **1-11**. Drilling ⅛" pilot holes for the screws lessens the risk of splitting the wood. Before continuing, make sure that each joint is square.

You'll need a leg in the rear corner of each of the main framework sections. I made mine from a 2 x 4 that I had on hand, but a 2 x 2 would suffice. Cut each leg about two inches

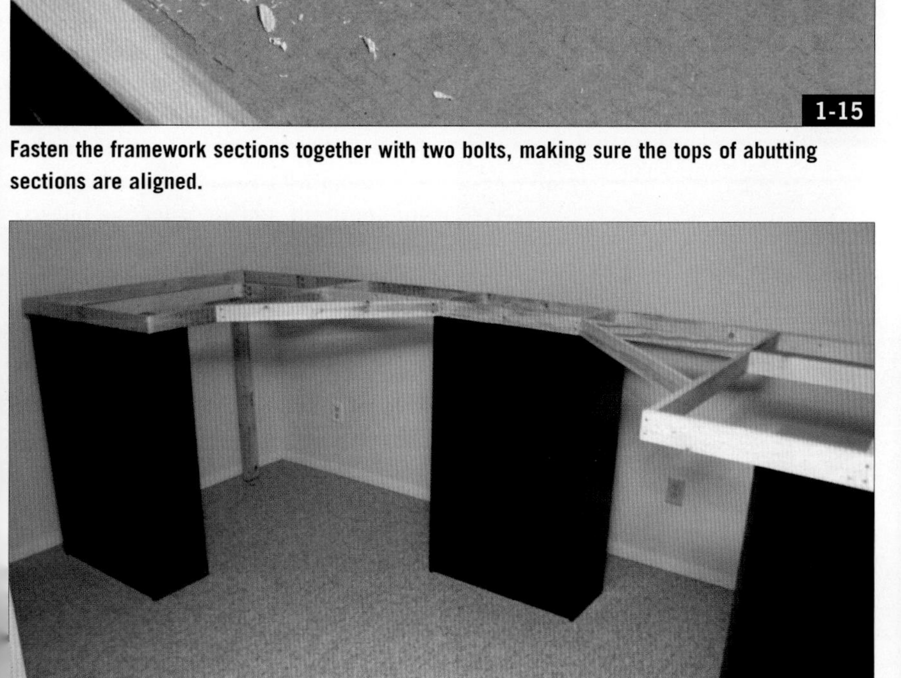

1-16

Once all three sections are together, adjust the legs so the benchwork is level.

1-17

Attach an L-shaped bracket at the rear of each front bookcase and screw it to the frame.

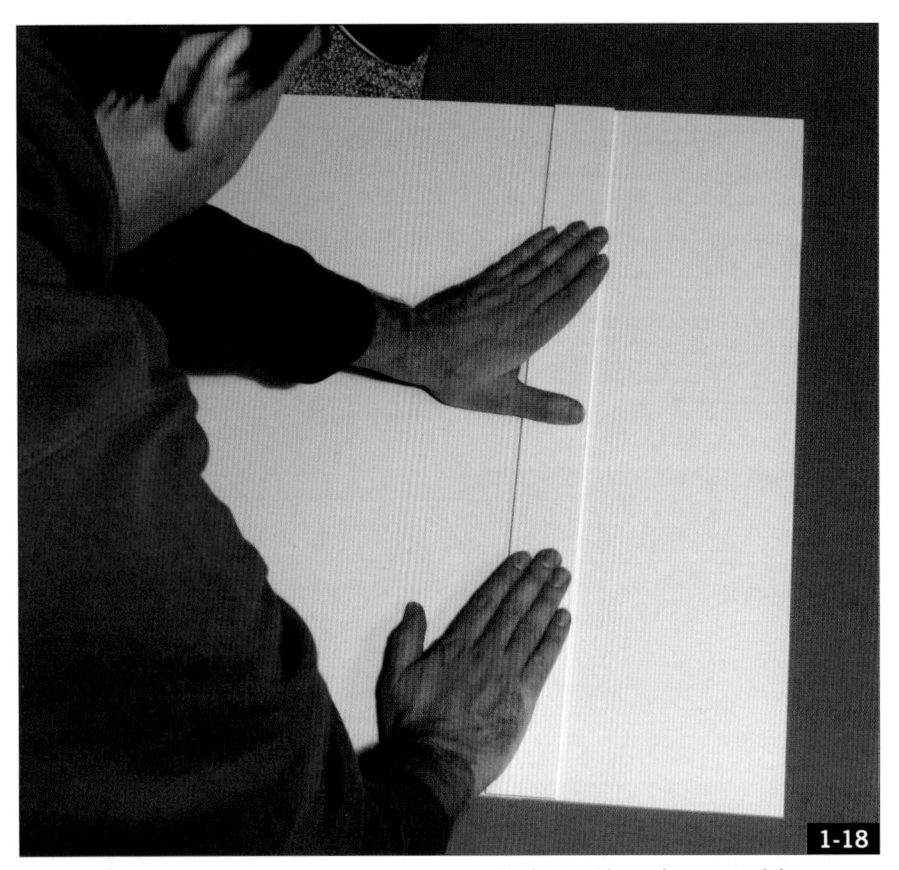

Butt backdrop sections against each other and glue a 2"-wide backing strip over the joint.

Apply Flex Paste or another filler to each joint and smooth it with a putty knife.

longer than the height of your bookcase, 46" in my case. The height isn't critical, as we'll make the legs adjustable by adding a T-nut and an adjustable foot.

To install the feet, drill a hole to clear the bolt in the center of the bottom of the leg, **1-12**. Screw the T-nut into place and add the feet. (These are available at larger hardware stores and home centers.) Clamp the leg into place in the frame and secure it with wood screws, **1-13**. If you want the ability to move the layout in the future, don't glue the leg in place. This layout isn't portable in the sense that it can be easily disassembled and moved multiple times. However, with a minimum of work and disruption, it is possible to take the layout apart and move it if necessary.

Place the bookcases in the proper spots, which can be seen in photos **1-14** and **1-16**. I covered each bookcase with a piece of cardboard to keep the framework from scratching or damaging the bookcase top. With the first benchwork section in place, the front edge of the layout framework should extend just beyond the front edge of the bookcase top, **1-14**, so the fascia will hang over the edge of the bookcase.

Add the middle section and other end frame, clamping each into place. Make sure the tops of the framework sections are aligned. Fasten the framework sections together with ¼" carriage bolts. Drill a pair of ¼" holes through the adjoining 1 x 3s, push the bolts into position, add a washer and nut to each, and tighten them securely with a wrench, **1-15**.

Make any final position adjustments to the frames and bookcases. Adjust the height of the legs so that the layout is level, **1-16**. My floor was fairly level; if your floor is uneven, you may need to add wood shims under the bookcases to keep them level.

Gravity does most of the work in holding the framework in place. But for extra support, I added an L-shaped steel bracket to each of the two forward bookcases, **1-17**. Place the bracket between a framework joist and the rear of the bookcase. Mark the hole locations, drill pilot holes, and drive in wood screws. If you add a backdrop, you may want to wait with this step in case you need to move the framework.

Backdrop

A backdrop isn't a necessity, but it certainly goes a long way toward making a layout more realistic. Backdrops – even plain blue skyboards – eliminate background distractions and allow viewers to focus on the layout itself. Also, if you are interested in taking realistic photographs of your model railroad, a backdrop makes it much easier. If you want to add a backdrop, the best time to do so is now, before the layout is built and scenicked.

You can use almost any flat sheet material as a backdrop. Thin plywood or hardboard works well but requires power tools for cutting. Large sheets of styrene plastic are also good, but they need framework to avoid sagging. You can use thin (½" or 1") extruded foam, but the surface often isn't smooth, and the softness of the material makes it vulnerable to dings and marks.

Sand the joint smooth and reapply filler if necessary.

Lengths of dimensional lumber or strips of plywood, screwed in place, hold the backdrop securely.

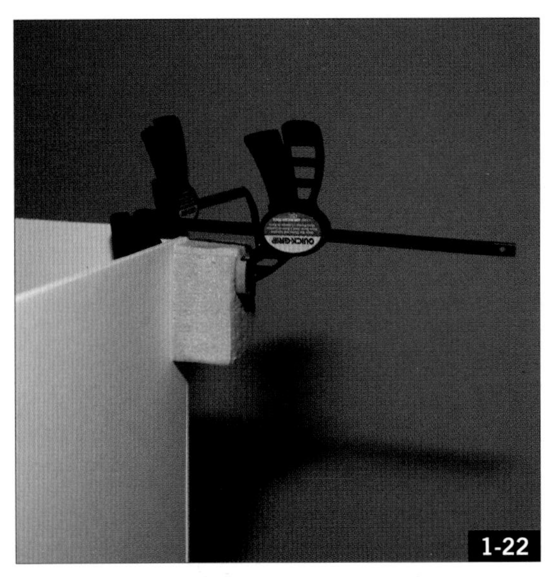

1-22

A piece of foam cut in an L shape secures the corner. Glue the foam in place and clamp it until the glue sets.

1-23

You may need a second set of hands to get the long rear backdrop section in position.

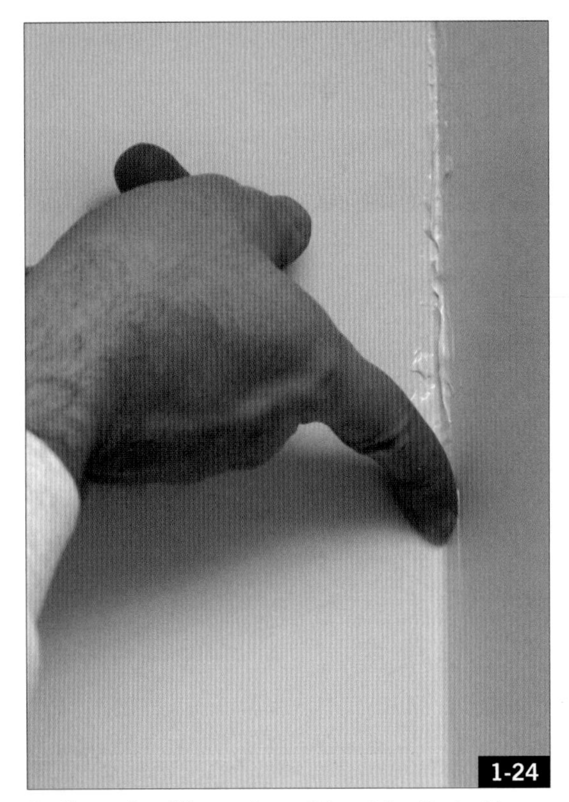

1-24

Caulk or other filler works well for giving the inside corners a clean look.

1-25

A vertical 1 x 3, glued in place, stabilizes the free ends of the backdrop.

For this freestanding layout, I used foam core board, which is available in art supply stores and some hobby shops. The material, foam sandwiched between paper, is easy to cut and takes paint well. Thicker foam core (at least ¼") stands on its own with minimal support. I placed several 20 x 30-inch sheets horizontally and glued the pieces together prior to attaching them to the benchwork.

Working on a large flat surface, butt two pieces of foam core end to end, and then splice them with thin foam core strips (about 3" wide), **1-18**. Spread Woodland Scenics Foam Tack Glue on the splice piece and press it into place over the joint. Add some weight (books work well) to hold it securely until the glue sets. Take care not to use too much pressure, as the foam core can become dented.

Repeat the process with successive pieces of foam core until the backdrop is the correct length. The backdrop section for each end has two pieces and the rear backdrop contains five.

When the glue is set, flip the backdrop over. You can hide any gaps in the butt joints with spackle or Woodland Scenics Flex Paste or Foam Putty, **1-19**. Apply the filler to the joint and smooth it with a putty knife. When it dries, sand the joint with a drywall sanding block or fine sandpaper, **1-20**. Add a second coat of filler if necessary.

Starting with an end, clamp the backdrop sections into place on the framework and then screw lengths of 1 x 2s or 1 x 3s along the bottom to hold the backdrops, **1-21**.

To make the joints secure where the end sections butt against the back piece, I glued the joints together. First, cut scrap foam into L-shaped pieces. Then, run a bead of glue along the end of the foam core and glue and clamp the L-shaped foam pieces in place, **1-22**. Add the remaining end piece and glue the other L joint, **1-23**.

Once the glue dries, smooth the inside of each corner. You can use paintable latex caulk, running a bead down the joint and smoothing it with a finger, **1-24**, or you can use spackle or Foam Putty. The goal is to soften the hard edge of the corner, which makes it more difficult to see.

1-26

Use a roller to paint the backdrop with sky blue latex and paint the corners with a brush.

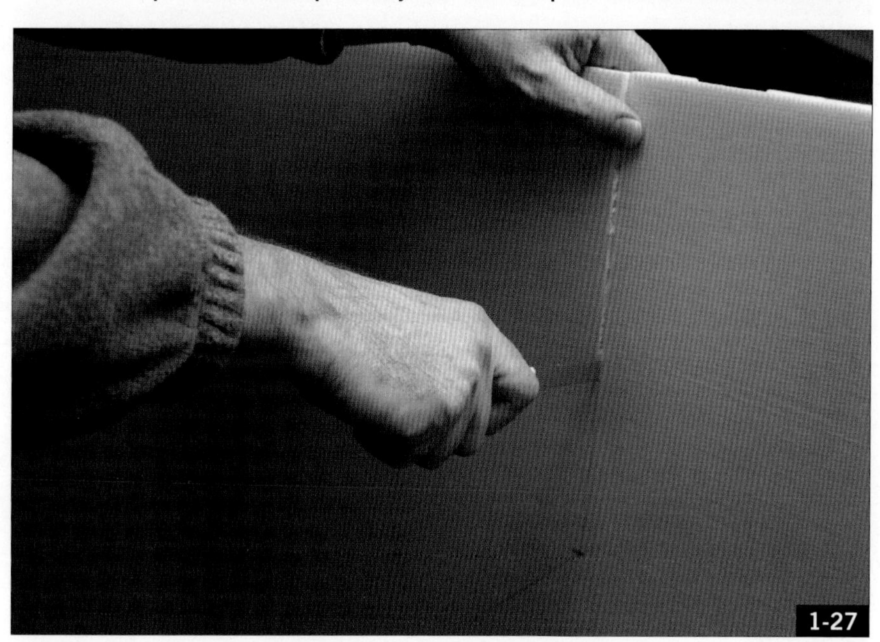
1-27

A serrated steak knife can be used to cut sheets of extruded foam.

Add a vertical 1 x 3 to protect each end of the backdrop, **1-25**. As the photo shows, I did it after painting the backdrop, but by doing this before painting, you will limit warping of the foam core.

Paint the backdrop a light sky blue color. Some modelers wait to do this, but you'll be happier if you do it now, before any track or scenery is in the way. Almost any brand of interior flat latex paint will do, and many shades of blue will work. (I used Glidden Evermore Delft China.) Get some color

1-28

Foam Tack Glue works well for gluing extruded foam to the wood framework.

1-29

Use clamps and weights to hold the foam securely to the frame until the glue sets.

sample cards and look at them under your room's lighting to determine a color that will look good. Blues that are too dark tend to be overpowering – stick with a light blue and make sure that it has a dead flat finish. Gloss and satin finishes will reflect too much light, giving the backdrop an unrealistic appearance.

Use a roller to apply the paint, **1-26**, making sure you completely cover the surface. A small brush will get the corners where the roller doesn't reach. A second coat is always a good idea for achieving a smooth, uniform finish. To avoid warping the foam core, give the back side of the backdrop a coat of paint as soon as you're done painting the front.

Spread glue on the first layer of foam, and then lay the top piece in place.

Layout table

The layout table itself is made from sheets of extruded polystyrene foam sold as insulation board at home improvement stores. It's available in several thicknesses, usually from ½" to 2½" in half-inch increments, and it comes in several colors (blue and pink are the most common), depending upon the manufacturer.

You have several options in using the foam. If you want to keep things simple and build a relatively flat layout, with no deep-cut features below the layout surface, you can use a single sheet of 1½" or 2" foam. On half the layout, I wanted to add a railroad bridge over a road and a road on a hill, so I placed a 2" piece atop a 1" piece.

Trim the edge of the foam so that it's even with the edge of the frame.

Making the table is simple. To start, measure the foam, draw cutting lines with a pencil or marker, and cut it to shape with a serrated knife, **1-27**. Next, glue the extruded foam board into place atop the framework. I used Foam Tack Glue, which adheres well to most types of foam as well as wood. Run a bead of glue along the exposed 1 x 3s on one framework section, **1-28**. Position the layer of foam, making sure the sides are aligned properly with the edges of the framework.

Add weights to the foam to hold it securely to the framework until the glue sets, **1-29**. I used heavy books and cans of paint, but almost anything heavy will work. Repeat this for each section and don't worry if there are small gaps between the sections.

Once the glue sets, you can add a second layer of foam if you like. Before doing so, read pages 55-58 on shaping and carving contours in foam. If you're doing extensive grade elevations and excavating major portions of the foam, you might find it easier to carve and shape the top layer of foam before gluing it to the base layer.

To add a second layer, spread glue on the first layer of foam, **1-30**, set the top layer of foam in place, and weight it down as with the first layer. Keep the weights in place at least overnight. Once the glue dries, trim the edges of the foam to match the edge of the framework, **1-31**.

The benchwork is about done. All that's left to do is add the fascia, the facing trim around the foam and frame edges. If you're planning some variations in scenery height with the foam, you'll want to wait and add the fascia after the basic scenery contour is done (pages 59-60).

Now, let's take a look at trackwork and wiring.

2-1

CHAPTER TWO

Track and wiring

Smooth-flowing, well-laid track is a key to ensuring good operations.

A model railroad that doesn't run properly will quickly become a dust collector, regardless of how good it looks. This makes trackwork arguably the most important step in building a layout, 2-1. Scenery, structures, and other features can be changed, fixed, and upgraded after a layout is finished, but track should be installed properly on the first try, as modifying and fixing mistakes can be difficult. It is important to take your time when laying track and to make sure everything works perfectly before proceeding to scenery.

Track basics

Most track consists of brass or nickel-silver rail held in place on plastic tie strips. Track sections are joined physically and electrically by metal rail joiners that connect the rails end to end. Sectional track, as the name implies, comes in various lengths and curvatures that can be combined to form any number of track layouts. Sectional track, although restricted by the curve radii available, is quicker to assemble than flextrack, which can be curved to almost any configuration but must be cut to length.

Traditionally, the most popular method of placing track onto a permanent layout is to lay the track (sectional or flex) atop cork, add loose ballast, and glue it in place. This results in very nice-looking track, but the process takes some time and may be more complicated than you want to attempt.

Combination track and roadbed, also known as all-in-one sectional track, is now widely available, and it is virtually ready to run. Several manufacturers offer lines of this track, which include Atlas True-Track, Bachmann E-Z Track, Kato Unitrack, and Life-Like Power-Loc, **2-2**. All-in-one track consists of sections having rails molded directly into the roadbed, such as Power-Loc, or sectional track pieces placed into ballast-textured plastic roadbed sections, such as True-Track.

Unlike standard track, each type of roadbed track uses its own style of connectors, so track from different manufacturers cannot be combined. However, these roadbed connectors do a better job of securing the track sections to each other than do rail joiners alone. Although most combination track doesn't quite match the appearance of ballasted standard track, especially when combined with scenery, it looks convincing. And as you gain confidence in your modeling skills, you can apply ballast over all-in-one track at a later time if you wish.

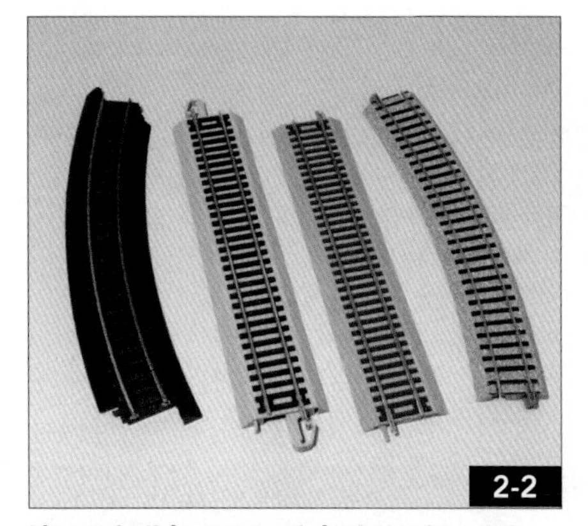

2-2 Lines of all-in-one track include, from left, Life-Like Power-Loc, Bachmann E-Z Track, Atlas True-Track, and Kato Unitrack.

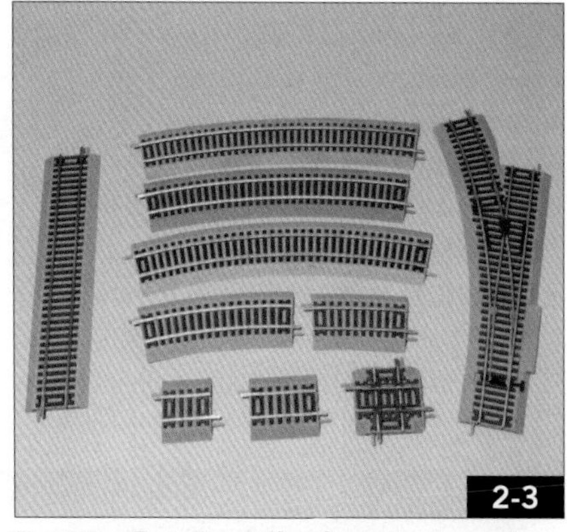

2-3 The Atlas True-Track line includes turnouts, a crossing, curves (from top) in 24", 22", and 18" radii, and assorted straight sections.

2-4 Test-fit the track on the table to make sure everything fits as planned. Run some trains to get a feel for how the layout will operate.

All-in-one track allows you to easily experiment with various track arrangements before you permanently secure it to the table.

I used Atlas True-Track for the layout, but other brands of all-in-one work just as well. The True-Track line, as do the others, includes a variety of components: crossings, left- and right-hand turnouts, curves in several radii, and a variety of straight and curved sections, **2-3**.

As mentioned in chapter 1, it's wise to start with a published track plan and modify it to suit your needs and interests. With the table completed, you'll now be able to see exactly how much track will truly fit into a given space – it is often less than it would appear on paper.

Along with standard full track sections, it's handy to have a variety of filler pieces on hand, such as half and third curve sections and straight sections in 6", 3", 2", 1½", and 1" lengths. These help you make fine adjustments when fitting track around buildings and other details.

In photo **2-4**, you'll notice that the track arrangement I'm putting together doesn't match the photos of the finished area or the track list, **2-5**. That's okay – now's the time to discover what works and what doesn't work. You'll also want to place structures and other scenic elements loosely in place to see how they fit with the track plan (page 8).

As you assemble the track pieces, make sure that all track joints are solid. Ensure that the roadbed connectors snap securely together and that each rail slides properly into its rail joiner. If the rail rides on top of the joiner, **2-6**, trains are certain to derail at that spot.

Track list

Unmarked curves are 18"-radius full sections

Unmarked straight sections are 9" long

Item no.	Description	Needed
450	9" straight	27
451	6" straight	4
452	3" straight	3
453	1½" straight	2
454	2" straight	6
460	18"-radius full curve	15
461	18"-radius ½ curve	4
*	18"-radius ⅓ curve	4
478	LH Snap-Switch	3
479	RH Snap-Switch	3

* from Snap-Switch or cut from no. 460 curve

2-5

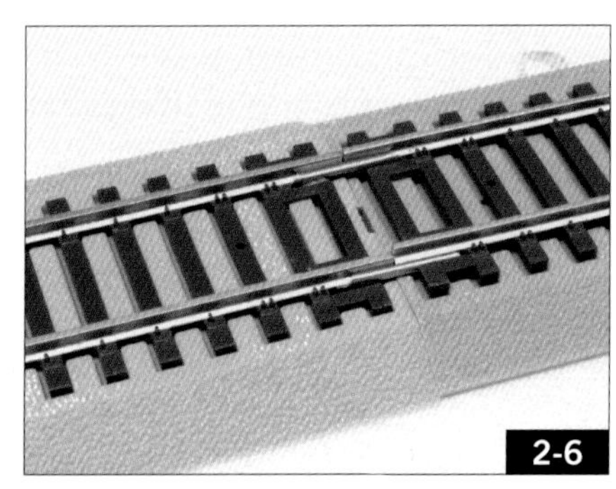

2-6 Make sure the rails slide into the rail joiners. Joints like this, where the rail is atop the joiner, will cause derailments.

2-7 Mark the track location by outlining it on the foam with a pencil.

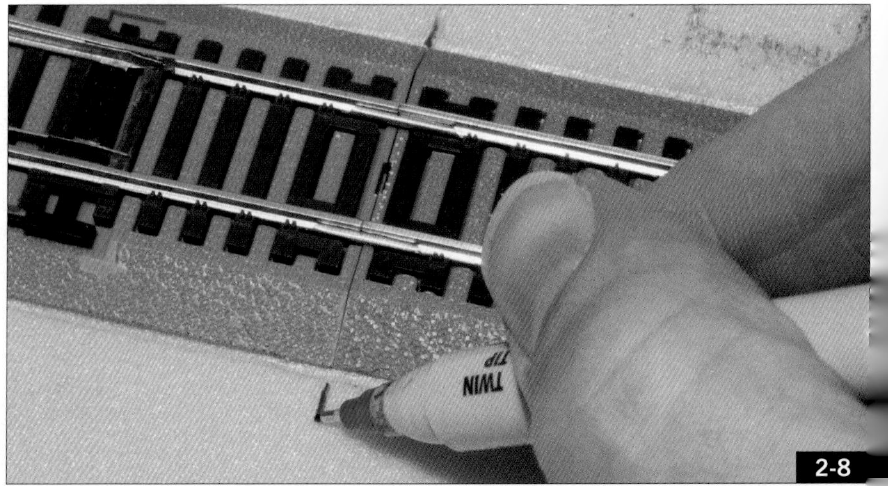

2-8 Although it is better to use a pencil for outlining track, since it is easy to get ink on the roadbed when outlining, you can use a marker to show the ends of a turnout.

Don't force track pieces into alignment or into tighter curves than designed. Instead, use small filler pieces as needed. If you discover you need a piece of track in an odd length, or one that you don't have on hand, you can cut a piece to size. On pages 22-23, you'll see how to cut custom track pieces.

Once you have the track together, take a few days to test your plan. Hook up your power pack or Digital Command Control system and spend some time just running trains. You'll quickly find out if this track arrangement is what you want. It's much easier to make changes now, rather than when the track is laid and scenery is in place.

When you're sure that the final plan is in place, mark the outline of the roadbed on the foam table with a pencil, **2-7**. Also add marks at the ends of turnouts, **2-8**, and between track pieces at a few other locations. These will serve as reference points to help you precisely place the track when it's time to glue it down.

Sketch your final track plan on a piece of paper, noting the location of the various track sections. Then remove the track, keeping it together in subassemblies. The sketch and the marks on the layout will help you correctly replace the track later.

Laying track

The proper time to lay the track depends upon the complexity of the scenery. I laid the track on the left side of the layout, which features only a few basic scenery contours, right away. However, if you are doing more extensive scenery work, as I did on the right half of the layout, you'll find it easier (and neater) to wait until you're done contouring the foam before laying the track (see chapter 4).

Glue, such as Woodland Scenics Foam Tack Glue, works well in securing all-in-one track to foam. Following the pencil-drawn outline marks as your guide, run a bead of glue down each side of the track, and another down the middle, **2-9**, and use a putty knife to spread the glue evenly, **2-10**. I find I can glue about three or four sections of track at a time.

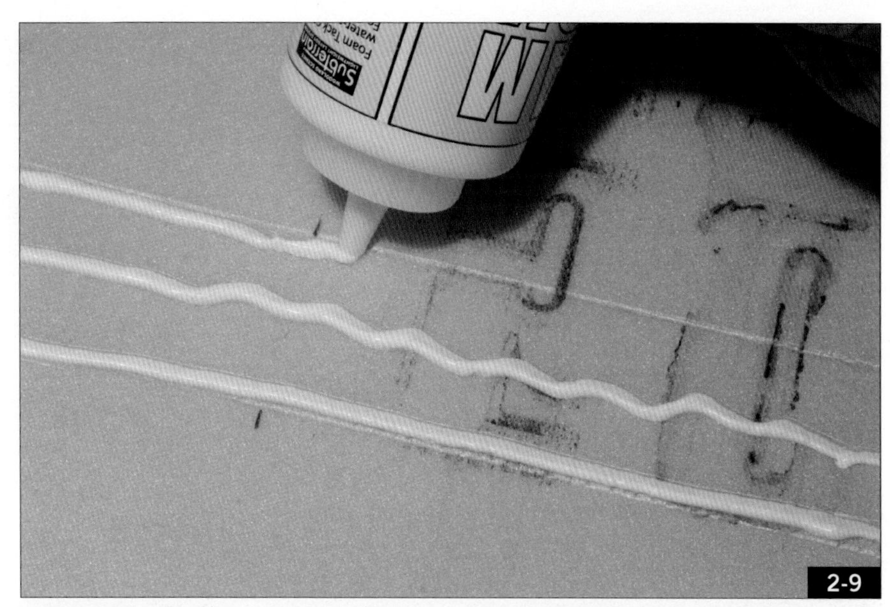

Run a bead of glue down each side of the track outline and another down the middle.

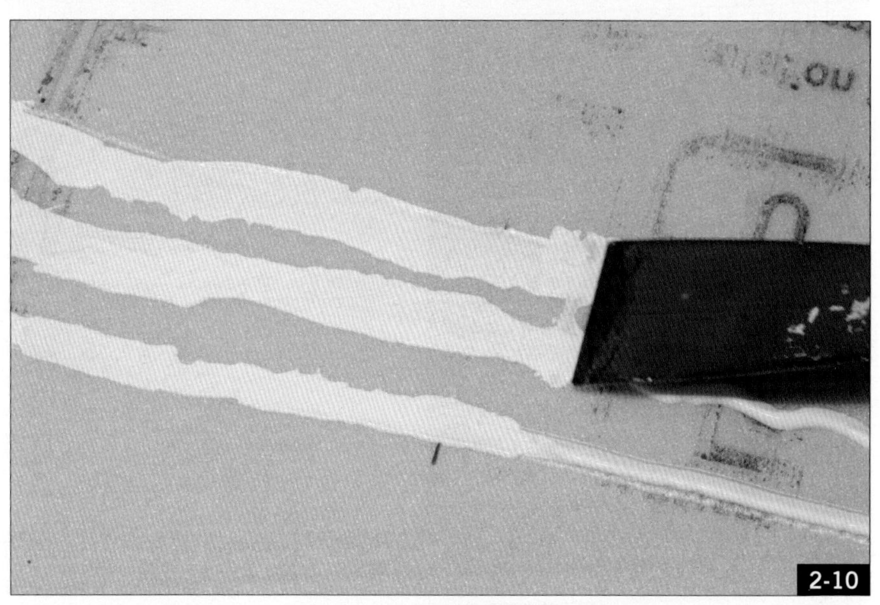

Spread the glue along the track area with a putty knife.

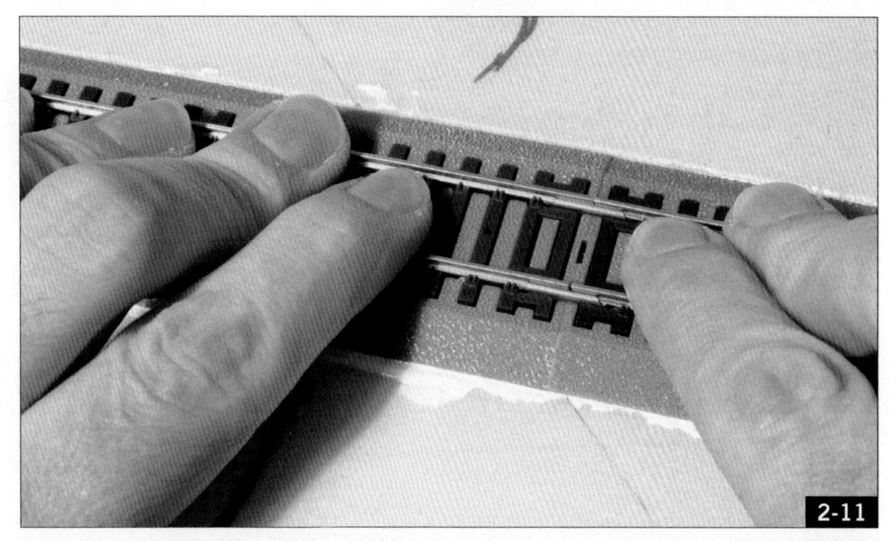

Press the track firmly into the glue, making sure the track stays within the marked outlines.

Custom track sections

Occasionally, you'll need a track section for which a commercial piece is not available or you don't have on hand. If this happens, don't panic – simply cut your own from an available piece of track. I used Atlas True-Track, so the following techniques may differ slightly for other brands.

Start by marking the location of the desired cut on the rails and roadbed, 1. Pop off the track and, at the marks, cut the roadbed with a razor saw, 2. Use rail nippers or a razor saw to cut the rails and tie strip at the appropriate location, 3. A flat file will smooth the ends of the rails, 4.

You'll need to remove the end tie from the cut end of the track to clear a rail joiner, 5. Save the tie. Use a hobby knife to remove the molded spike detail, recess it slightly under the rails, and place it into the appropriate tie slot in the roadbed strip. The rail joiners will now slide in place over it.

Use these modified pieces the same as any other track section. Although the roadbed won't have the locking connectors, the rail joiners will keep the track aligned, and it will be secure once glued in place on the layout.

Modifying turnouts is a bit trickier but still not too difficult. Start by cutting the roadbed at the rail joint between the turnout and the ⅓ section of curved track, 6. You'll then have to trim the edges of the adjoining track sections slightly, 7, so they don't interfere with each other, 8. With the track in place, the roadbed modifications are hardly noticeable, 9. As the photo shows, instead of using two ⅓ curve sections, you can cut your own ⅔ curve section to eliminate one rail joint.

1

Mark the location of the cut on the rails and roadbed with a fine marker.

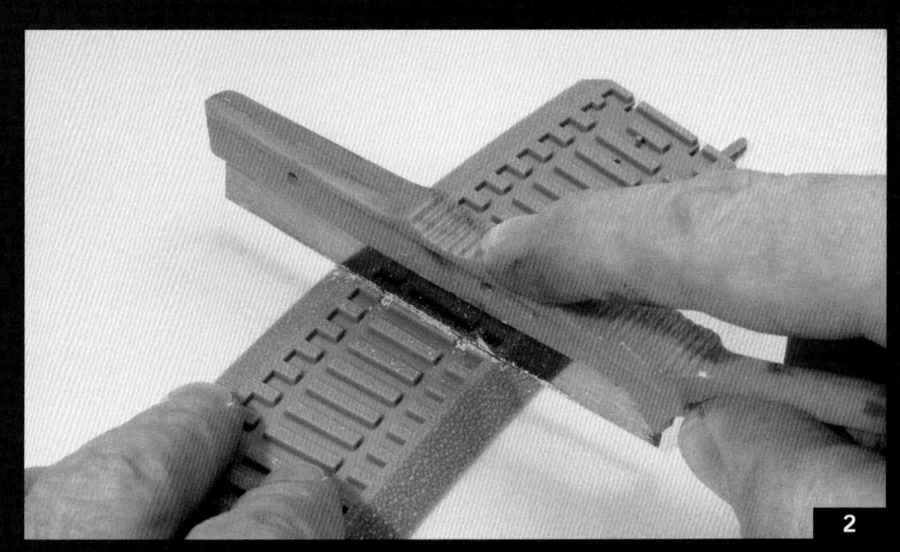

2

Cut the roadbed using a razor saw. Make sure the cut is square and vertical.

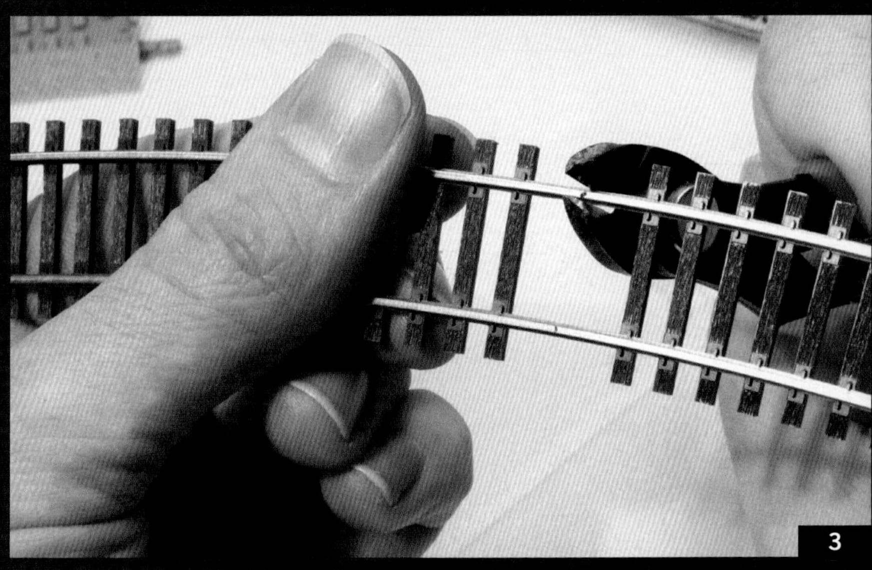

3

Cut the rails with a rail nipper or razor saw.

File the cut end of the rail to ensure a smooth connection with the rail joiner.

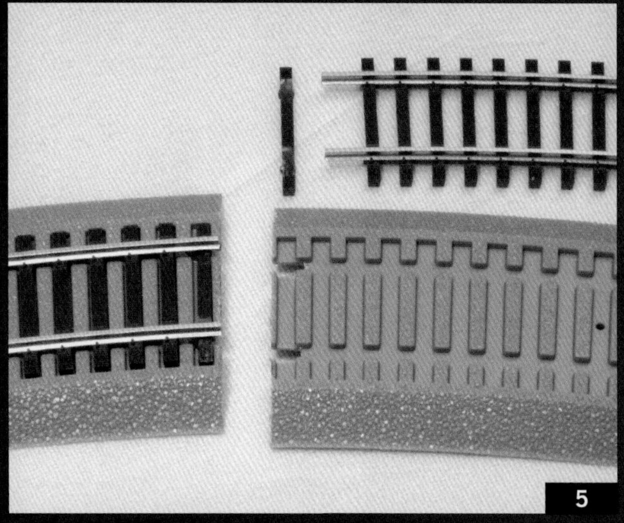

Notch the loose tie to clear the rail joiners and set it in place and put the track back on the roadbed.

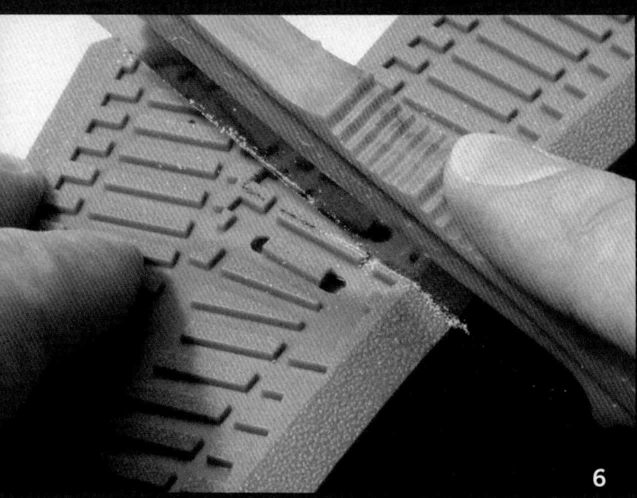

To modify a turnout, cut the roadbed on the diverging route at the rail joints.

Use a pencil to mark the adjoining roadbed sections that need to be trimmed. The pencil marks can easily be erased.

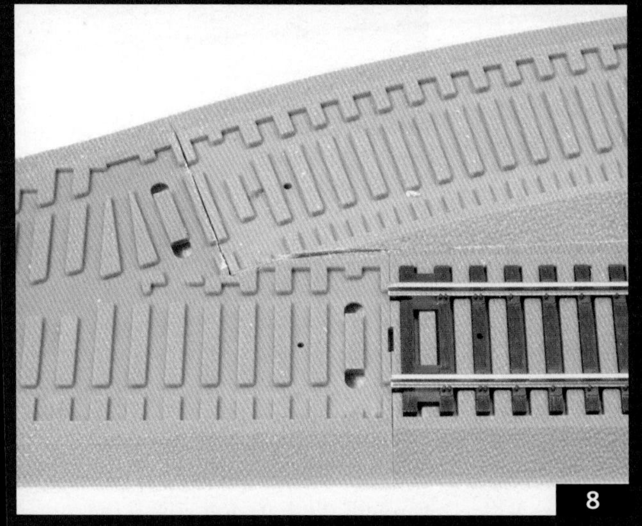

Test-fit and trim the roadbed sections until they fit snugly and then add the track.

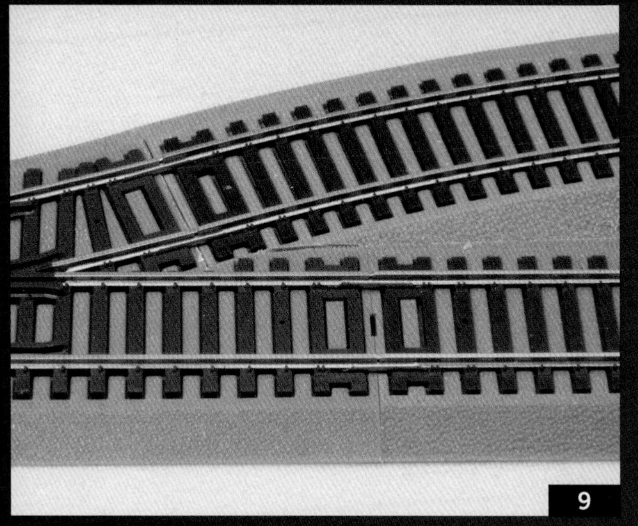

When the track is in place, the roadbed joints are not readily visible.

Place weights atop the track to make sure it bonds firmly to the foam.

Rail joiners with presoldered wires make it easy to get power to the rails.

Carve notches in the roadbed with a hobby knife to clear the wired joiners.

Double-check that all the roadbed connectors are firmly in place and that the rails are securely joined and aligned properly. Set the track into position, following the markings, and press each piece firmly to the foam, **2-11**. Placing weights atop the track will hold it until the glue sets, **2-12**.

To provide for power connections, you'll need to add track feeders while laying track. The easiest method is to use terminal joiners, which are rail joiners with wires presoldered to the bottom (Atlas no. 553), **2-13**. When adding terminal joiners, you may have to enlarge the gaps at the ends of the roadbed section with a hobby knife to clear enough room for the wires to pass, **2-14**.

Don't rely on a single pair of feeder wires to power the whole layout. Add terminal joiners every five or six track sections to ensure solid electrical contact throughout the layout.

When you determine a location for the terminal joiners, drill a hole through the foam using a screwdriver blade, **2-15**. Mark one wire, with a bit of paint or a marker, as the inside rail to ensure that all wires will be the proper polarity when they're connected later. Thread the wires through the hole and attach the joiners between the track sections and glue them into place as with the other track.

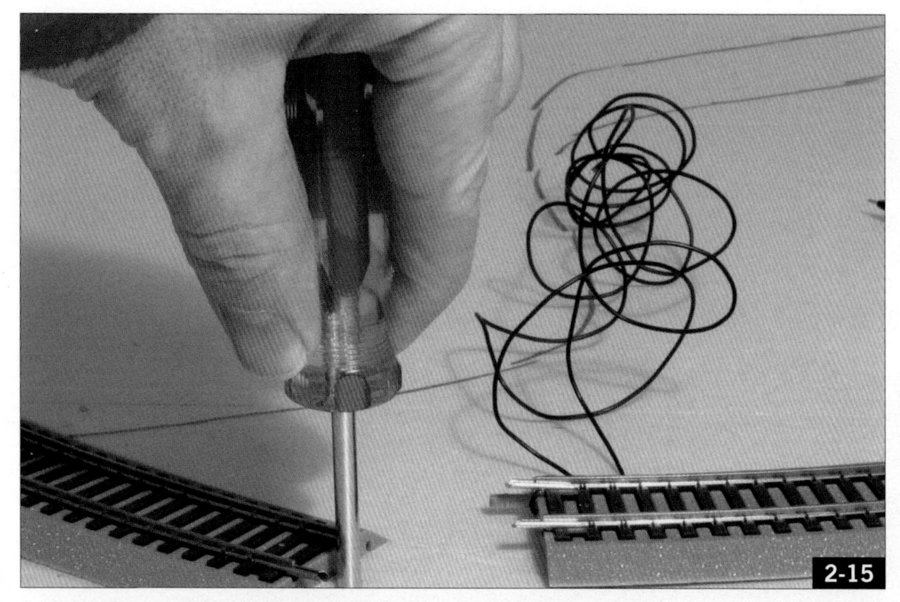

A screwdriver blade is handy for drilling holes through the foam table.

Floquil paint markers work well for painting the sides of rail a rusty brown color.

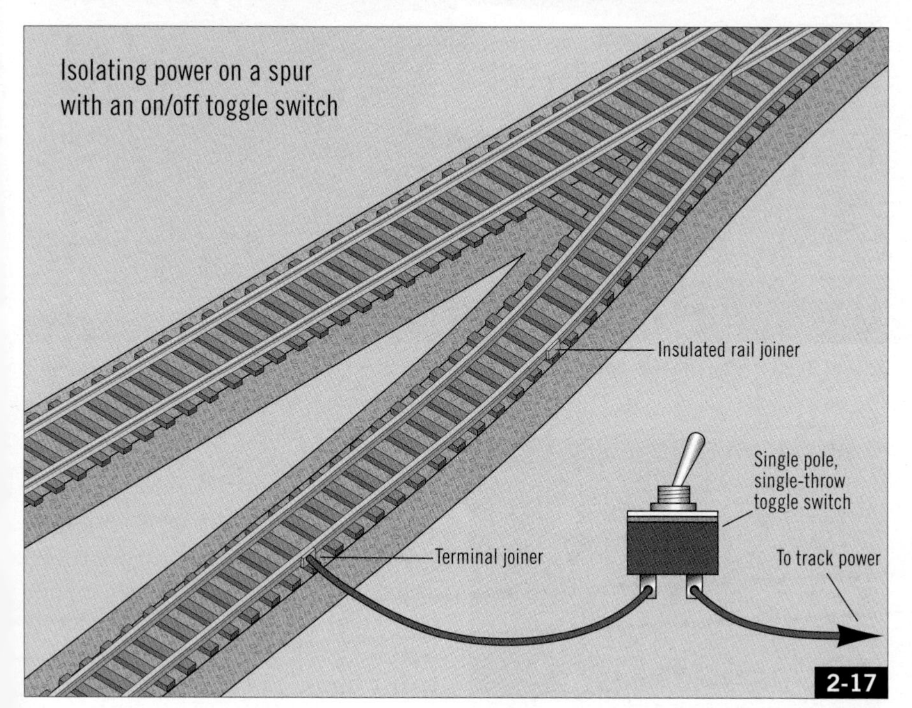

This layout is wired as a single electrical block and will be powered by a small DCC system. Even if you want to use a standard DC controller, I'd recommend keeping it wired as a single block since it would be difficult to operate two trains at once and keep them out of each other's way. However, you could store a locomotive on a stub track (or two) by wiring it with a simple on/off switch, **2-17**.

Turnouts

Turnouts are available in manual and remote-control versions. I chose manual turnouts, as the layout is small and all the turnouts are within easy reach.

The two turnouts at the ends of the passing siding required some modification to get the track spacing that I needed. The Atlas turnouts include a ⅓ curve section on the diverging route, which makes it impossible to make a siding with closely spaced tracks. (I discovered this after having all the components on hand.) To make a siding that will parallel the main line, this ⅓ section needs to curve in the other direction. (Other brands include the ⅓ section as a separate piece that can be used in either direction.)

I solved the problem by cutting each turnout to fit, as shown on pages 22-23. Other solutions include modifying the track plan or using another brand of track that allows this track arrangement without modification.

Painting rail

One unrealistic aspect of model track is that the nickel-silver rail is bright silver and shiny, whereas, the sides of real rail features a myriad of rust colors. Painting the rail on model track goes a long way toward improving its appearance. Painting rail before the track is in place on the layout is risky, as rail joiners often won't make good electrical contact with painted rail. This is why it's wise to paint the rail after the track is laid.

The handiest products I've found for doing this are Floquil's enamel paint markers, which are available in various colors. I used markers from weathering set no. F3801 (Rail Brown, Rail Tie Brown, and Rust) to paint the rail sides, **2-16**. The process goes rather quickly but be sure to completely cover the plastic spike heads. You might need to apply two coats for complete coverage. You can also mix different colors to achieve the streaked, multihued look of real rail.

Some paint will inevitably get on the tops of the rail, which can interfere with electrical contact and cause rough operation. Clean this off as soon as the paint dries by rubbing the railheads

The Digitrax Zephyr (on the shelf) is a Digital Command Control combination command station and throttle. The UT-4 throttle (hanging on the fascia) is a handheld unit that plugs into the Zephyr.

A notch in the rear of the bookcase allows wires to pass through from the command station to the layout.

with an abrasive track cleaning block such as a Bright Boy.

If you have a flat layout, the track is now done. My layout has several special areas (the highway overpass and a spur over a coal trestle) that require some prep work before tracklaying can proceed. Chapter 3 explains how to take care of these and other special situations.

Wiring

You have two options for controlling a train on this layout: a standard DC power pack or a Digital Command Control (DCC) system. If you already have a train set or an earlier layout that used a power pack, you can use it again for this layout. If you're just getting into the hobby, consider starting with DCC. Digital Command Control allows you to control multiple locomotives on a layout without the need for separate electrical blocks or boundaries. The DCC command station sends signals through the rails, which are picked up by decoders mounted inside each locomotive.

Although not a necessity for a small layout like this one, which is wired to operate one train, you'll find having a DCC system to be quite an advantage if you expand the layout or plan to build a larger one in the future. Along with running multiple trains, DCC systems provide greater control over features such as locomotive lighting (headlights, ditch lights, and warning beacons) and sound systems. Decoder-based sound units are extensive, allowing control of engine sounds as well as horns, whistles, bells, coupler crashes, and many other sounds that can increase the realism and operational value of a layout.

Several affordable entry-level systems include the Atlas Commander, Bachmann E-Z Command, Digitrax Zephyr, and MRC Prodigy Express. Many manufacturers offer locomotives, steam and diesel, with factory-installed DCC decoders and sound units. Other locomotives are DCC ready, meaning they generally have a built-in socket that accepts a plug-equipped decoder.

For this layout, I chose a Digitrax Zephyr that features a throttle built into a command station, **2-18**. Getting this entry-level DCC system up and running is as simple as plugging it in and connecting two wires to the track.

I placed it on the top shelf of the middle bookcase. Cutting a small notch in the rear panel of the bookcase with a hobby knife allows the wiring to pass through, **2-19**.

Wiring the layout is simple: Just connect the two wires from the DCC system or power pack to the feeder wires that were attached to the track earlier, **2-20**. Make sure all inside rails are connected and then check the outside rails.

Most DCC-equipped locomotives come from the factory preset to address 03. To run one of these locomotive, place it on the track, turn the track power on, select locomotive 03 on the command station, and turn the throttle dial up. The keypad allows you to turn the headlight on and off, and if you have a sound-equipped model, you can use the various function keys to actuate the whistle or horn, bell, and other sounds.

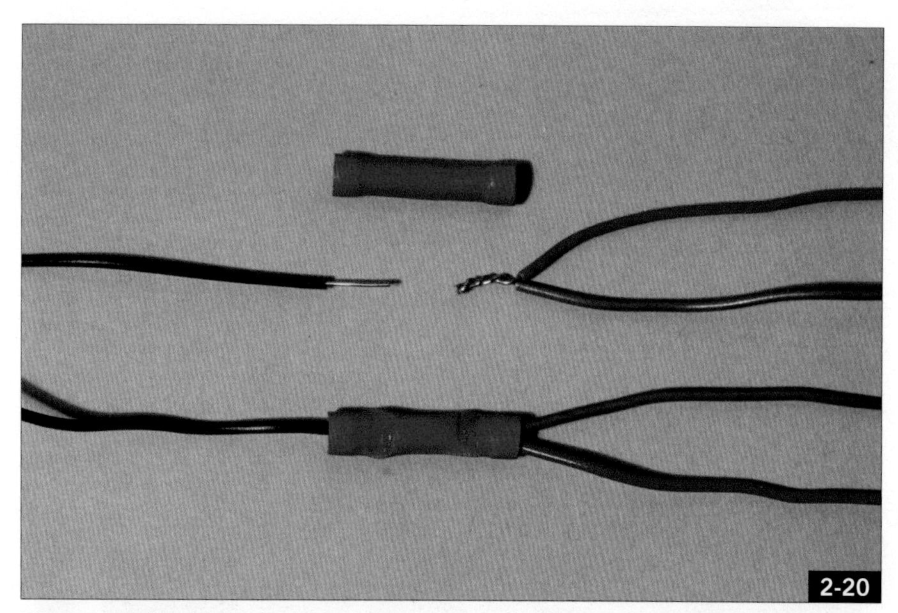
2-20

When connecting feeder wires, wire joiners can be crimped to secure the ends of wires at joints. Twist the wires together before crimping.

2-21

Velcro on the fascia and the back of the throttle allow it to hang out of the way when not in use.

You will find detailed operating information in the user manual for your system. There are many options and adjustments available, but the basics of running a train are easy right from the start.

I added a Digitrax UT-4 handheld throttle to my Zephyr. It simply plugs into the rear of the Zephyr and hangs on the fascia on the front of the layout when not in use, **2-21**. Although strictly optional, the handheld throttle makes it easy to follow your train around the layout when switching cars or throwing turnouts.

Let's continue turning the table into a model railroad with a look at structures.

3-1

CHAPTER THREE

Structures

Structures give a layout personality, and there are many ways to personalize ready-built or kit structures to make them unique to your layout.

Storefronts, houses, industrial buildings, and other structures give a layout life and purpose. Railroads have destinations for loads and people have places to go. Buildings also help define a layout's region and era and create interest for viewers. Today, modelers have access to many types of structures as either factory-assembled models or easy-to-build kits. And if you like, these buildings – with a few easy improvements – can be personalized to create unique structures for your layout, 3-1.

3-2
Several basic assembled structures are available, such as Barb's Bungalow from Atlas and a yard office from Walthers.

Your choice

Atlas, Walthers, Woodland Scenics, and other companies offer assembled structures ranging from basic bungalows, **3-2**, to highly detailed and weathered storefronts, **3-3**. The selection of assembled structures is limited when compared to the number of kits available, so you should consider building some kits to complement your assembled selections. I used several of each type on the layout, **3-4**.

Don't feel bound to use the same structures as I did. You might want to choose structures that resemble those of a prototype town, make up a specific scene, fit an available space, or just catch your eye. You can always upgrade or change structures later.

Easy improvements

One problem with many ready-built structures and some plastic kits is that their empty windows give them a vacant, unoccupied look. This is especially true with storefront buildings. You don't need to add full interior detailing to make a model look lived in. Simply by adding a few signs, window coverings, and other details, you will greatly improve the appearance, **3-5**.

Let's look at two representative storefront structures, Melissa's Eastside Deli and the Silver Dollar Cafe from Walthers. Straight from the box, they're nice models of typical storefronts that can be found along main streets throughout the country. Both buildings include several styles of hanging sign frames and sheets of peel-and-stick signs for a variety of businesses. However, instead of using the signs that came with the buildings, I wanted to personalize my models.

A wide variety of signs that can be placed in windows or on the structures themselves are available, **3-6**. Paper signs are made by Blair Line, City Classics, JL Innovative Design, and others. Microscale has decal sets for steam-era and contemporary signs, and dry transfer signs are available from Clover House and Woodland Scenics. Three-dimensional structure and billboard signs are offered by Bar Mills and Blair Line, and Miller Engineering offers a variety of illuminated signs.

To customize a building, first select a business for each structure. I decided to make one a drugstore, **3-7**, and the other a paint store, using signs from Blair Line. Start with the hanging signs. The sign frames that come with the kits are ready to mount in holes already in the wall. This is what I did with the Rexall sign, which I cut from a Microscale decal sheet and mounted.

For the paint store, I cut out the hanging signs from a Blair Line set and mounted them to a styrene sign board.

Window signs make the building appear active, and they also hide the lack of interior detail by blocking the view inside. I used paper signs from Blair, JL Innovative, City Classics, and others. Glue them in place by smearing the faces of the signs with a clear-parts adhesive (I used Model Master) and pressing them into place on the

3-3
The Corner Emporium and Harrison's Hardware from Woodland Scenics are examples of assembled structures that come with signs and much detail.

Kits on the layout

1. Woodland Scenics Clyde & Dale's Barrel Factory (BR5026)
2. Walthers water tower (933-2826)
3. Walthers rural grain elevator kit (933-3036)
4. Rix Products corrugated grain bin kits (305)
5. City Classics gas station kit (108)
6. Walthers Melissa's Eastside Deli (933-2815)
7. Walthers Silver Dollar Cafe (933-2816)
8. Design Preservation Models Robert's Dry Goods kit (102)
9. Walthers Water Street building (933-2814)
10. Walthers Golden Valley depot (933-2806)
11. Walthers O.L. King & Sons Coal Yard kit (933-3015)
12. Walthers Wallschlager Motors (931-805)
13. Woodland Scenics Corner Emporium (BR5024)
14. Woodland Scenics Harrison's Hardware (BR5022)
15. Design Preservation Models factory kit (built from several modular component kits)

3-4

inside of the clear window glazing. The glue dries clear, so only the sign is visible.

You can also make your own window, hanging, or structure signs as explained on pages 32-33. You'll also learn how to make signs that appear to have lettering painted directly on windows.

Window shades add character and also block the view to a building's interior. Most upstairs windows in the buildings received window shades, **3-8**. These are easily made by cutting paper and thin cardstock of various colors and gluing it in place behind the windows, **3-9**. Vary the colors and the shade positions for variety.

Another trick to keep viewers from looking directly through a structure is to add a view block of black construction paper, **3-10**. One piece generally does the trick; if a structure can be viewed from more than one angle, you might need to add a second piece.

3-5

Melissa's Eastside Deli building (Walthers) was upgraded, at left, with a few signs, window shades, and some light weathering.

If a building with large windows is located near the edge of a layout, where it's easy to see inside, you can take the idea of a view block to another level. I did this for the paint store. Cut a piece of heavy cardstock or black matboard to fit as an interior wall and glue a photo of a store interior to it, **3-11**. Using my computer, I scanned a generic photo of some store shelves,

Printed signs are available from City Classics (left), Blair Line (center top and bottom), and JL Innovative Design (bottom right); decals (upper right) are made by Microscale.

Walthers Silver Dollar Cafe became a drugstore courtesy of a number of paper signs in the windows and a Blair Line Rexall building sign.

Window shades make a building look occupied, and they keep viewers from seeing a lack of interior detail.

Signs on the computer

1

The State Bank sign and clock were made using a photograph taken of a real sign.

If you have a computer and a digital camera (or a scanner), you can model any prototype sign you can photograph. Most cameras and computers now come with basic photo manipulation software; more advanced programs such as Adobe Photoshop and Photoshop Elements allow additional operations. As an example, I'll take you through the steps I used to make the bank clock sign, 1, that I added to the corner bank building.

Start by loading the image into the software. Using a flatbed scanner, I scanned a photo that I had taken and opened the image in Photoshop Elements, 2. Size the image to fit your sign. For the best quality, make sure the image's resolution is at least 300 dpi at the size you plan to use it.

Make any color adjustments and image corrections. If you photograph a sign at an upward angle, the sign will have a bit of keystoning, with the bottom of the sign appearing wider than the top. There might also be side-to-side distortion if the sign was photographed from the side. These are easy fixes in Photoshop Elements. Go to the Image tab, Transform, and then Distort. You'll be able to grab the corners of the image and pull them into the proper shape.

Then make any other modifications you desire. I removed the Cuba City lettering to make it a more generic State Bank sign.

2

Photoshop Elements and other imaging software allows you to resize, edit, adjust colors, combine elements, and otherwise modify photos and artwork for use as structure signs.

Print the sign on glossy photo paper for the best appearance. To save on cost, you can print several signs on one sheet.

Trim the sign and affix it to a heavy (.060" or .080") piece of styrene using double-sided photo mounting paper, 3. This double-sided sticky paper is easier to control than glue and works instantly. Trim the styrene to match the shape of the sign, 4, and add the sign on the opposite side.

I mounted the sign to the building with two pieces of .028" brass wire. With a pin vise, drill a pair of no. 69 holes in the edge of the sign, 5, and glue a length of wire in each, 6. Drill a pair of matching holes in the structure for mounting the sign.

The Baumann Fuel Co. sign, 7, was done the same way, but it started with a logo from a matchbook cover that I found in an antique store. Because the logo on the matchbook was nice and square, most of the modifications involved resizing it to fit a hanging signs from a Walthers storefront kit. I cleaned up a few stains and wrinkles from the matchbook and printed it out. As the photo shows, I printed it in several sizes for signs placed on vehicles as well as the building itself.

Window signs are also easy to make using drawing and painting sofware. I made the signs in the bank building windows with the AppleWorks drawing feature on my Mac, 9. The window sign is simply white lettering on a black background. When glued inside the window, it looks like lettering that's painted on glass with a darkened room behind it. The sign doubles as a view block, keeping viewers from seeing through the windows.

The insurance company sign was done the same way, but it uses a photo of a curtain as the background instead of black.

Keep an eye open for potential sign sources, including matchbooks, stationery, books, and promotional items. Antique stores are a great source, as are online auction and reference sites.

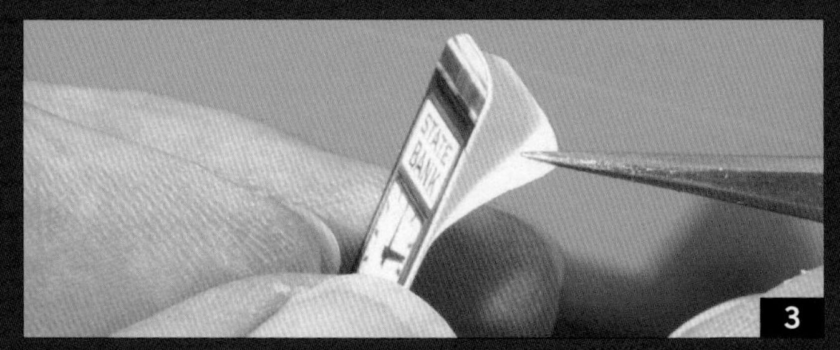

Double-sided mounting paper is excellent for securing the sign to its base. With a tweezers, peel off the backing sheet to reveal the adhesive.

Trim the styrene with a hobby knife to match the shape of the printed sign.

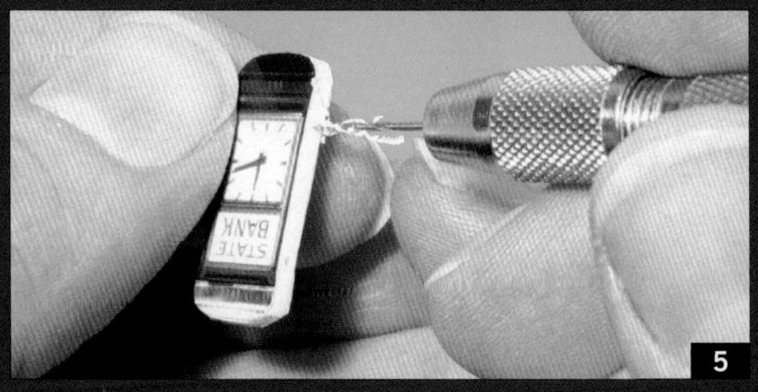

Use a pin vise to drill mounting holes in the side of the sign to match the size of the mounting wires.

After the sign is complete, paint the wires dark brown and add the sign to the building.

The finished Baumann sign looks sharp on the coal dealer building.

The fuel company signs, based on a matchbook cover, were sized to fit a Walthers sign board.

These window signs were printed on photo paper and glued inside the window glazing.

Signs are easy to make, and they let you customize structures to match specific prototypes or even immortalize friends. You can use these techniques with hanging signs, storefront windows, billboards, road signs, and other signs. Templates of the signs shown here can be found on page 36 for your use. *[A note on copyright: Most logos of current corporations and businesses are trademarks of their respective companies. You have the right to photograph signs and use these items to make model signs for your own use. But unless you obtain permission, you generally do not have the right to distribute them to others, regardless of whether you charge for them or not.]*

reduced it to the proper size, and printed it. I glued it to matboard and then glued the matboard to several pieces of styrene strip and positioned it about 12 scale feet inside the front of the building. The shelves don't have to be highly detailed or even accurate for the type of store you're modeling – just so there's something there to give the impression that the store has an interior.

You can also add details inside a storefront's windows, as I did with the grocery store shown in **3-12**. First, I made shelves of sheet styrene and glued them inside the windows and glued a few fruit crates from Preiser atop the shelves, **3-13**. Again, it's enough detail – coupled with the window signs – to give the impression of a detailed interior, even though the inside of the store is bare.

The Wallschlager Motors building from Walthers, **3-14**, has the large front windows typical of older car dealerships. The structure needed more than window signs (which come printed on the model) to hide its vacant interior. And because the building stands alone on the layout, I added some other details to it as well.

I started by building a simulated concrete base to fit around the building (more about bases on pages 38-39). Having a base also makes the building easier to place in scenery, as chapter 4 explains. The base is made from several pieces of Smalltown USA's injection-molded styrene sidewalk sections (no. 7000), glued side-by-side with plastic cement, **3-15**. The base is a scale 20 feet or so larger than the building in each direction, which leaves a sidewalk around the building (about 10 feet wide around the sides and front and about five feet on the back). I painted the base Polly Scale Concrete for the sidewalk area, and Maintenance of Way Gray for the interior floor, **3-16**.

To create a car showroom, I glued an automobile at an angle on the floor inside the front window. This doesn't have to be a top-of-the-line model – I used an old plastic Eko car that was lurking in my scrapbox and detailed it by painting its hubcaps and grille silver. I also added two Woodland Scenics

3-9

To create shades, cut pieces of colored paper or construction paper to shape and glue them behind the windows.

3-10

A black construction-paper view block keeps viewers from looking all the way through a structure.

3-11

Black matboard with a photo or artwork glued to it makes a convincing interior wall.

3-12

This grocery store looks occupied thanks to window signs and several fruit and vegetable crates resting inside the windows.

3-13

The crates are simply glued to styrene shelves, which are glued to the front walls just below the window bottoms.

Building signs

You can use these signs on your own layout by copying them, scanning them into a computer and printing them, or cutting them from the book (see pages 32-33).

FIRST STATE BANK

FIRST STATE BANK

FIRST STATE BANK

FUEL OIL COAL BAUMANN FUEL COKE CO.

FUEL OIL COAL BAUMANN FUEL COKE CO.

STATE BANK

STATE BANK

CAVEY INSURANCE AGENCY

The large windows of Wallschlager Motors called for more interior detail than did the storefront structures.

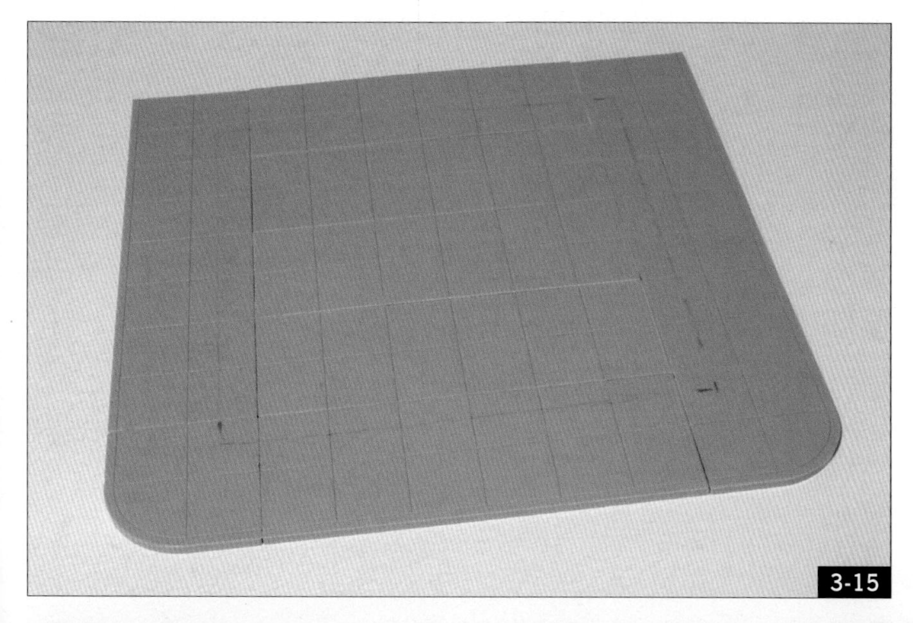

The base for the Wallschlager building is made from several Smalltown USA sidewalk sections glued together.

The base has been painted, and a car and two figures have been glued in place, so the building can be glued to the base.

Powdered chalks, such as these from AIM Products, make it easy to weather structures, locomotives, and rolling stock.

figures as a salesman and customer and a piece of black matboard as a wall between the showroom and the garage area.

When the details are in place, use cyanoacrylate adhesive (CA) to glue the building in place atop the base. The finished effect is quite convincing and gives the impression that the structure serves a real purpose on the layout.

Most prebuilt structures and plastic kits are molded in somewhat accurate colors, but they are not painted. However, bare plastic, regardless of its color, often has an unrealistic shine that doesn't replicate the look of an actual brick or a painted finish.

Lightly weathering the building with powdered chalk is a simple way to improve this condition. You can scrape artist's pastels with a hobby knife to make a powder, or you can buy chalks already in powder form, such as the weathering colors set from AIM Products, **3-17**.

Gray mortar colors work well to weather brick buildings. You can mix almost any shade of gray by combining white with black or grimy black. Use a stiff-bristle brush to work a bit of chalk into the surface, **3-18**. It may take some time to both kill the shine of the plastic and add some realistic variations in color.

For the best effect, use black and dark gray chalks around smokestacks and on roofs and rust-color chalk on metal siding and roofs and around detail items.

Structure bases

Many structures come with their own injection-molded bases. Whether or not to use them depends upon how you plan to use the buildings on your layout. Storefront buildings, such as the Walthers models described earlier, often have a base that includes a front sidewalk. When using a variety of structures, they tend to have different sizes of sidewalks. It's better to have a standard sidewalk, so I trimmed off this detail on my buildings, **3-19**. The rest of the base is still useful, as it can elevate a building to sidewalk height. On the drugstore building, it provided the portion of the sidewalk under the recessed doorway.

3-18

Use a stiff brush to work the chalk into the mortar lines of the brick.

3-19

The front sidewalk has been trimmed off the storefront structure's base at right.

On a free-standing building, if the base is easy to blend into the scenery, use the base. If it isn't, you can blend it directly to the layout or build your own base.

Some structures don't include bases, and sometimes they need one. Because the Woodland Scenics hardware store has a number of details attached to the front and side walls that are designed to rest on a sidewalk, I added a base to it using a piece of foam core, **3-20**. Since I planned to place this build-

A piece of foam core, painted a concrete color on the edges and top visible areas, serves as a base for this Woodland Scenics store.

The base supports all of the exterior structure details and makes it easy to plant the building into the scenery.

ing on a hill (more on that in the next chapter), I added a base that would serve as the structure's sidewalk and serve as a step from the streetside sidewalk that would be added later. Part of the edge of the base will be visible when the structure is planted, so I painted the edges and sidewalk areas Polly Scale Concrete. Then I glued the building to the base with CA, **3-21**.

Structure kits

Several buildings on the layout are built from kits, and even if you've never built a model before, I encourage you to assemble a basic plastic structure kit. This will improve your modeling skills and give you access to a far wider range of structures than if you limit yourself to assembled models. For detailed information on structure modeling, see *Basic Structure Modeling for Model Railroaders* (Kalmbach Books).

It's now time to give our structures a home. Let's move to the next chapter and get going on some basic scenery.

4-1

CHAPTER FOUR

Scenery foundations

Basic scenery includes grass and brush ground cover, trees, highways, farm fields, and other features.

Scenery is the most important step in transforming a train set on a table into a realistic model railroad. Of all the aspects of model railroading, scenery is probably the least "ready to run," which makes many modelers afraid to take the plunge into scenery. However, scenery is very forgiving – it is easy to redo areas that don't turn out quite right, and getting basic green texture over everything is easy if you take things step by step.

Hills, bridges, and other variations in elevation help improve realism by eliminating the tabletop appearance.

Scenic planning

First, determine the type of scenery you want to feature on your layout. Approaching Midwestern or Plains scenery is certainly different than modeling mountains or a Southwest desert. For this layout, I wanted to keep things fairly basic but have some varied features as well to make the layout more interesting. Overall, the final scenery is typical of the upper Midwest with generally level topography and some rolling hills and variations in elevations, **4-1**.

Building a layout on a flat tabletop is the easiest approach, especially if you want to get a layout scenicked in the shortest amount of time. However, adding scenic features both above and below the table level adds greatly to realism, giving the feeling that the railroad runs through the land, not just on top of it, **4-2**.

The left side of the layout features a small town on a level surface.

The town on the left side of the layout is built on the level table, with just a ditch or two cut into the foam surface, **4-3**. On the right side, I built the town on a hill and added a rail bridge over a road, a track on an embankment, and a coal yard that has a track on an elevated trestle, **4-4**. Note in the photos how much this adds to the realism of the scenery.

As mentioned in chapter 1, if you prefer, you can certainly keep both halves of the layout level. And you don't have to make your scenery exactly as I do – use the following techniques as a guide in designing your own scenery.

Remember, at this stage, the track location has been finalized and outlined, but the track in some areas has not yet been glued in place (see chapter 2).

The right side of the layout has a road on a grade, a bridge, and other variations in depth.

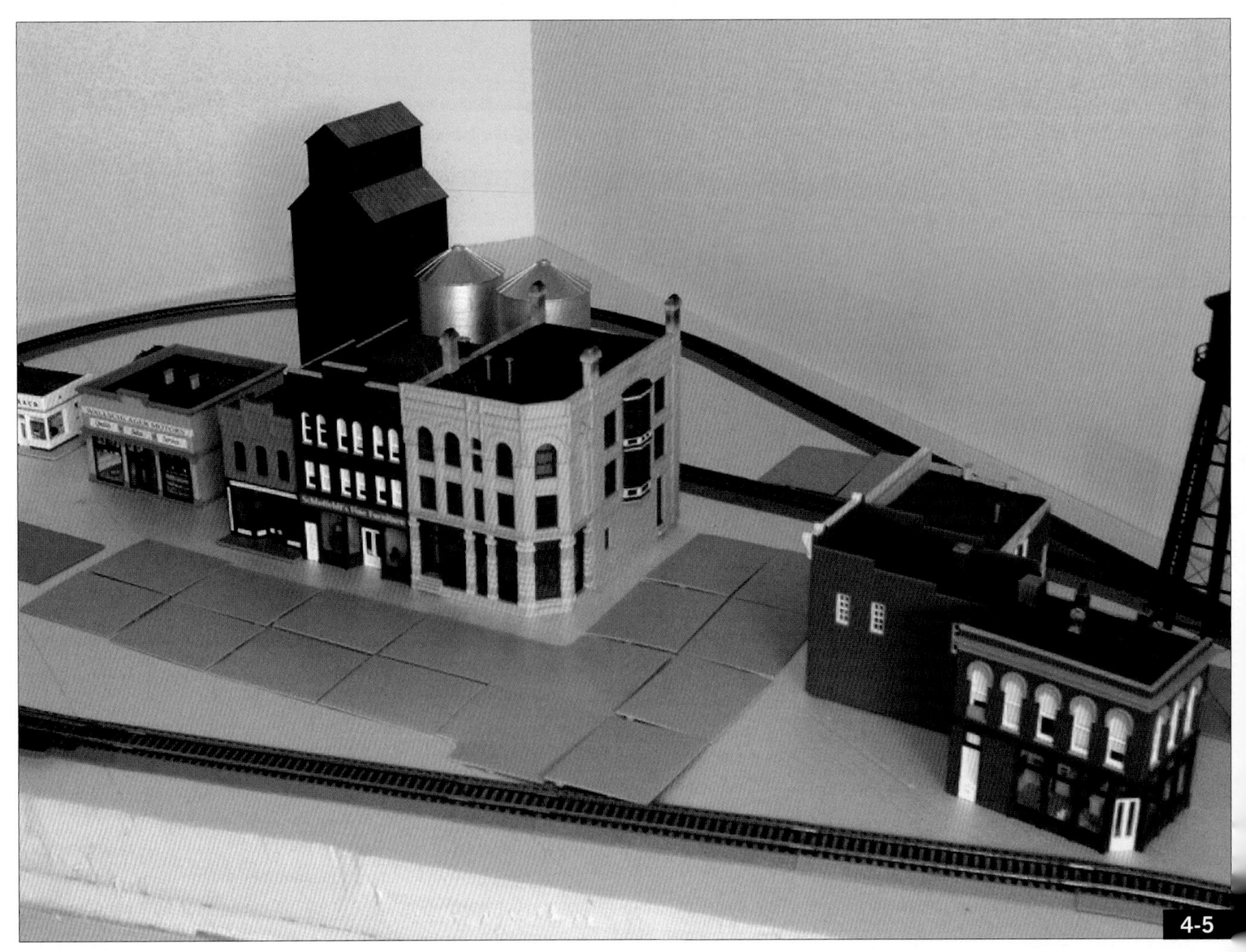

Experiment with various arrangements of structures and roads until you find the effect you like best.

Gather the structures and other scenic features you plan to use to make sure they fit on the layout and between the track, **4-5**. The best way to do this is to place these items on the layout and move them around until you find an arrangement that works well. Don't be afraid to deviate from drawn plans, as things often go together on the layout differently than they appeared on paper.

You can draw in features such as roads and structure locations on the foam top itself, or on sheets of drawing or craft paper, **4-6**. Sketching on paper makes it easier to redraw the plan multiple times if necessary. It took a few tries before I found an arrangement of structures, roads, and streets that I was happy with.

As noted in chapter 1, when I glued the foam layers together, I hadn't planned on excavating as much as I wound up doing. It was a case of not planning well enough in advance, but it also illustrates that it's relatively easy to change plans as you go along. If I had it to do over, I would have carved the top layer of foam before gluing it in place on the base layer.

You can draw features, such as streets and roads, on the foam itself or on large sheets of paper.

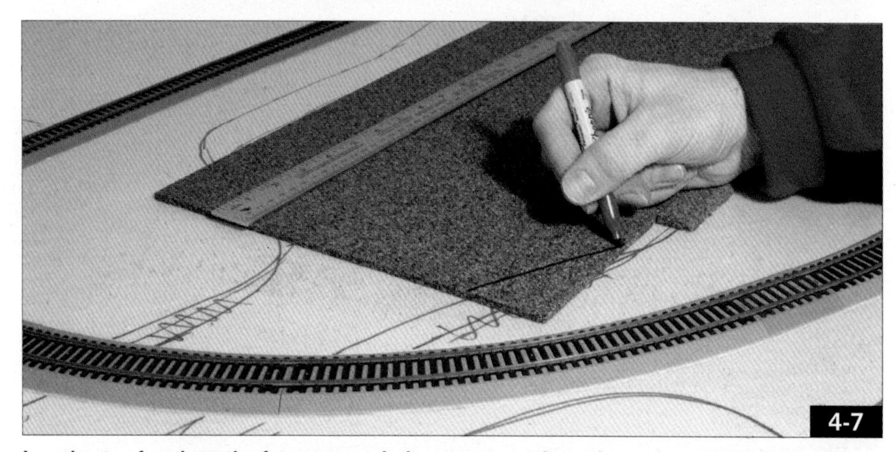

Lay sheets of cork on the foam to match the streets, roads, and storefront structures on the flat end of the layout. Use a hobby knife to cut the cork to fit.

Highways and roads

I used several methods to add roads to the layout. For the town on the left, I decided to use Walthers concrete street components (no. 933-3138). This modular system features various street, curb, and sidewalk sections that work well for urban and town scenes. The street itself comes in two halves, each of which includes a traffic and curb lane. The resulting street is also crowned, making it higher in the middle than at the gutters, as many real streets are.

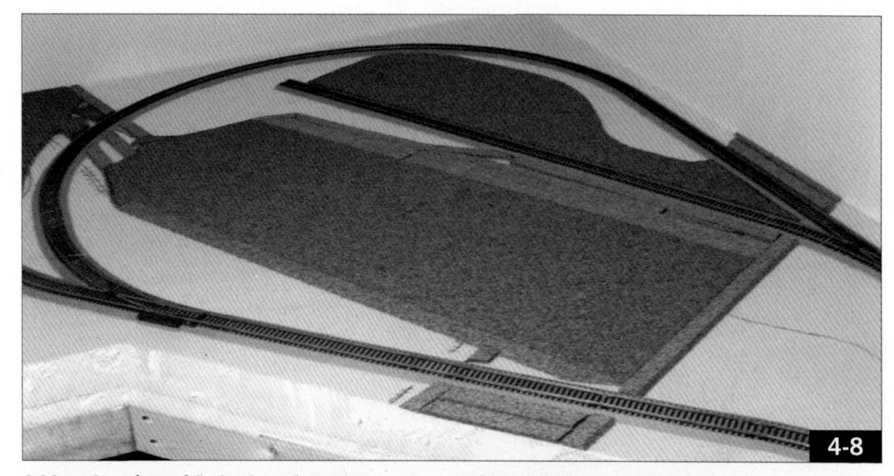

Add cork strips with the beveled edge on the outside around the cork sheets.

Glue the modular street sections to the cork with cyanoacrylate adhesive (CA) and make sure the pieces fit tightly together.

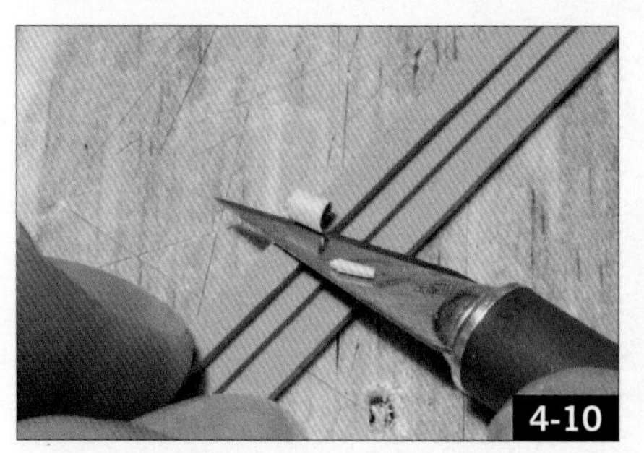

Trim each outer grade crossing segment with a knife until it sits level on the track.

Test-fit the pieces, add CA to the track, and press each grade crossing section into place.

Use a freight car to make sure the flangeways are clear through the crossing.

Mark the street section so that it will butt snugly against the grade crossing.

Start by cutting the pieces from their sprues. Arrange the pieces on the layout, along with any structures you plan to include in the scene. This will help determine where you should locate cross streets, railroad crossings, and other details. It took three concrete street kits to complete the scene. When you're happy with the arrangement, mark the outline of the pieces on the foam with a marker.

Roads, like railroads, are elevated from ground level to provide drainage. Duplicating this on a layout adds to realism by providing variation in scenic levels. Cork roadbed and sheets work well for raising roads (and also as bases for structures).

Large street areas, including surrounding structure footprints, are best cut from large cork sheets (available from Midwest Products), **4-7**. Cut the cork to shape with a hobby knife and use roadbed strips with their beveled edges on the outside, **4-8**. Glue the cork in place using Woodland Scenics Foam Tack Glue. Placing books and other weights atop the cork (as with the foam in chapter 1) will keep it flat while the glue dries.

Glue the modular street sections in place, **4-9**. Use liquid plastic cement to glue the pieces to each other, and medium-viscosity cyanoacrylate adhesive (CA) to glue them to the cork base.

As your streets and roads approach the tracks, you'll need to add grade crossings. Walthers makes injection-molded simulated rubber crossings (no. 933-31370) that work well with straight track. Each crossing has three pieces: one for the middle section between the rails and a pair of outside rail sections.

Start by cutting the pieces from their sprues. The outside sections may need a bit of trimming with a hobby knife, **4-10**, to fit properly with the tie and roadbed profile of the all-in-one track. Remove enough material so that the piece is level when in place. Paint the crossing components with Polly Scale grimy black and glue the pieces in place with CA, **4-11**. Double-check the clearance for the train's wheels by rolling a freight car through the crossing, **4-12**.

Once the crossing is in place, add the remaining street sections. Mark the street piece from the grade crossing kit, **4-13**, and score the line with a hobby knife, making a few light passes. A steel straightedge works well as a guide for the knife. Bend the piece at the scored line to break it, **4-14**. Clean up any stray fuzz on the cut edge with the knife and glue the piece in place, **4-15**.

Once the street sections are in place, add the two-piece curb and sidewalk sections, **4-16**. Glue each curb/sidewalk piece together. Start at a corner and work along the street, gluing them in place with CA until done. Cut them to fit as needed at grade crossings. I didn't use a sidewalk on the side of the street by the depot.

For the two-lane road heading out of town, I used Rix concrete road sections (no. 106), **4-17**. These pieces are the same as those found in the company's excellent bridge kits. The sections represent a typical concrete slab highway, with expansion joints and rough texture molded in place.

4-14

The best way to cut styrene is to score the line with a hobby knife and carefully bend the piece at the score until it snaps.

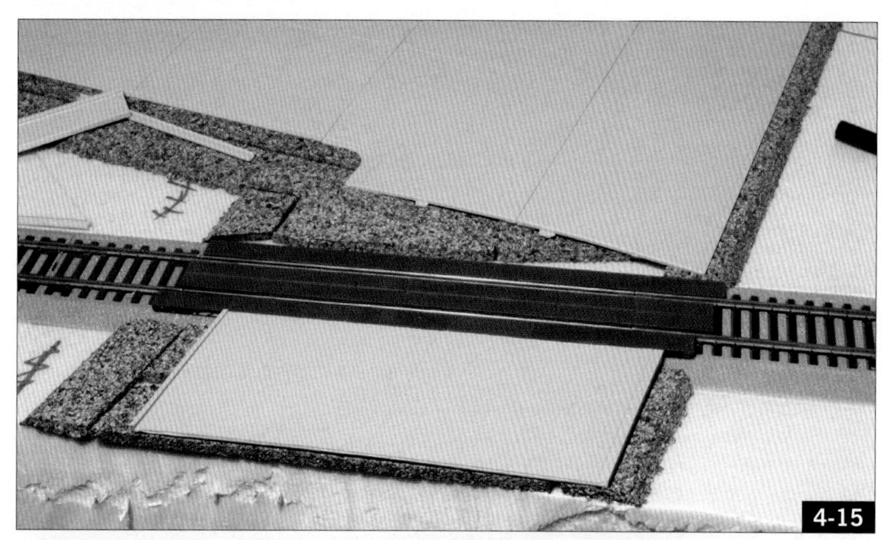

4-15

Continue gluing the street sections in place, filling in the rest of the gaps.

4-16

The sidewalk and curb are separate pieces. Glue them together and then to the street and cork base.

Cut the two-lane Rix road section to butt against the last full street sections and then cut other street sections to fit as needed.

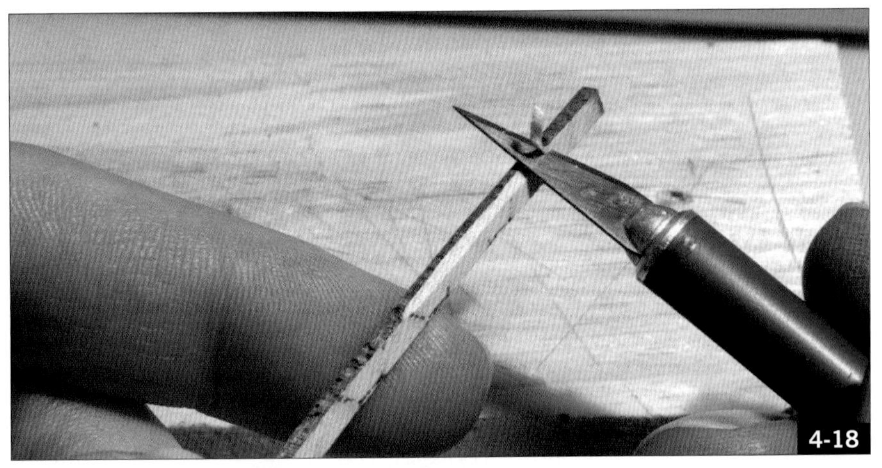

Trim the rail side of each outside grade crossing piece to clear the spike heads along the rail.

To blend the Rix and Walthers sections, I made the joint at a slight angle, which resulted in trimming the sctions. To do so, score and snap the Walthers and Rix pieces as needed to form the joint. The dimensions and angle aren't critical – cut the pieces as needed to fit your specific needs.

Wood plank grade crossings were common through the late 1900s, and some can still be found in use today. Blair Line makes nice laser-cut wood kits for straight and curved versions. I used a curved one where the Rix highway crosses the tracks.

The wood pieces that are placed on the outside of the rails must be trimmed slightly to clear the molded spike heads along the rails. This is easy to do with a hobby knife, **4-18**. Glue the pieces in place with CA. You may have to adjust the length of the crossing to match the width of the road – I had to make a longer one, so I combined two crossings. Simply cut the pieces as needed at the scribe marks and glue them together. Because the road sections are thinner than the wood grade crossing, I used .030"-thick styrene strips to shim up the road at the crossing, **4-19**. Glue the final road section to fit between the crossing and the backdrop.

You can use these components for the roads on the other half of the lay-

The completed Blair Line crossing is in place. The .030" styrene strips shim the road to the height of the planks.

out as well, but I wanted to use the right side to show how to build highways out of large sheets of styrene plastic. This is an advanced but versatile highway-building method that allows you to build roads and streets to any configuration.

Start by making a pattern for the road. As shown in photo **4-6,** I did this before contouring the foam in the city area by putting the structures in place and drawing the road outline on a large piece of brown craft paper. The road will be cut as a single piece, from the inside edge of the layout to the grade crossings at the rear and the right end.

Large pieces of styrene can be purchased at plastics dealers in most cities (check the phone directory under Plastics). I used .040"-thick material. Cut the road pattern from the paper, tape it to the plastic sheet, and trace its outline with a marker, **4-20**. Score the plastic with a hobby knife, freehanding the curves and using a steel rule to keep other lines straight, **4-21**. Bend the plastic at the scored lines and the material will snap cleanly. Test-fit the road section, **4-22**. Allow some extra length on each segment of the road, which will be trimmed to fit once the foam is contoured.

Sand the plastic to prep it for painting. I used a rubber sanding block with a piece of 120-grit sandpaper, going over the surface in a circular motion until the plastic shine was gone and the road had some texture.

If you're keeping this part of the layout flat, continue installing the road; if you're adding hills to the foam scenery as described in the next section, wait and install the road after the foam is contoured.

When the foam is ready, lay the road in place, and mark its location on the foam. Build up a road subbase using cork as with the concrete highway, **4-23**. Glue the cork in place with Foam Tack Glue and secure it with pins or weights.

You'll need to cut the road at the outside rail of each grade crossing. One way is to paint the railhead with a china marker or lipstick (asking permission first, of course!), **4-24**, lay the road in place, and press it over the rails.

Tape the pattern of the road onto the styrene and trace its outline with a marker.

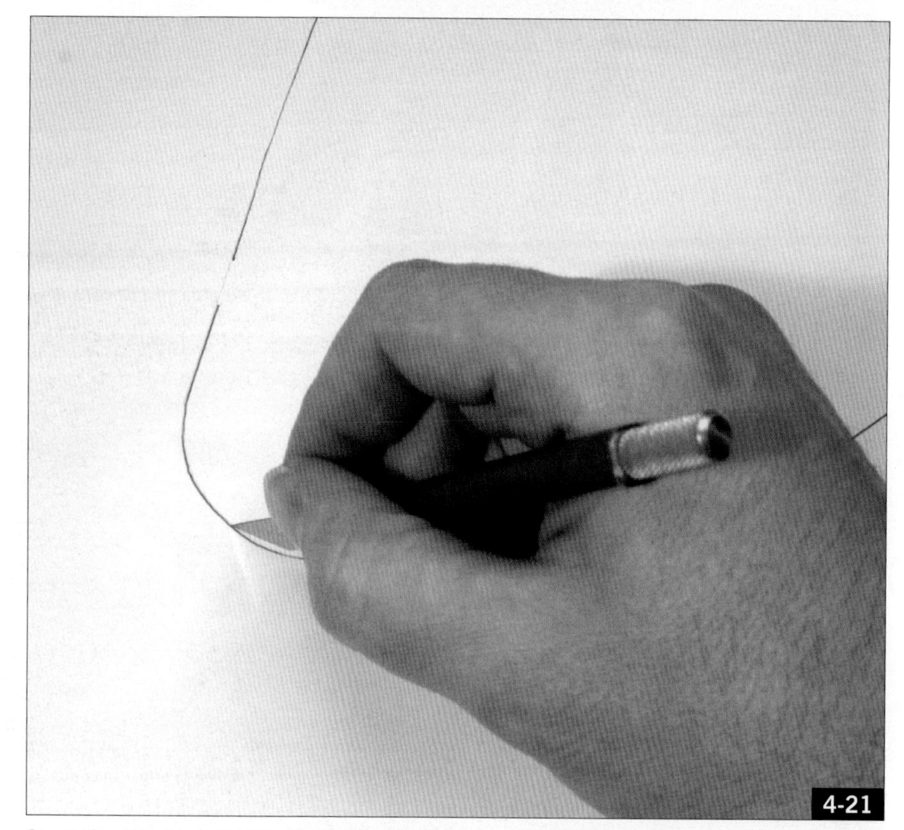

Score the styrene along the lines with a hobby knife. Freehand the curves but use a steel rule to guide the knife on straight sections.

Test-fit the road in place. You can see that the final structure arrangement differs from this test-fitting session.

Add cork roadbed as a base for the road and glue it in place. Pins or weights will secure it while the glue sets.

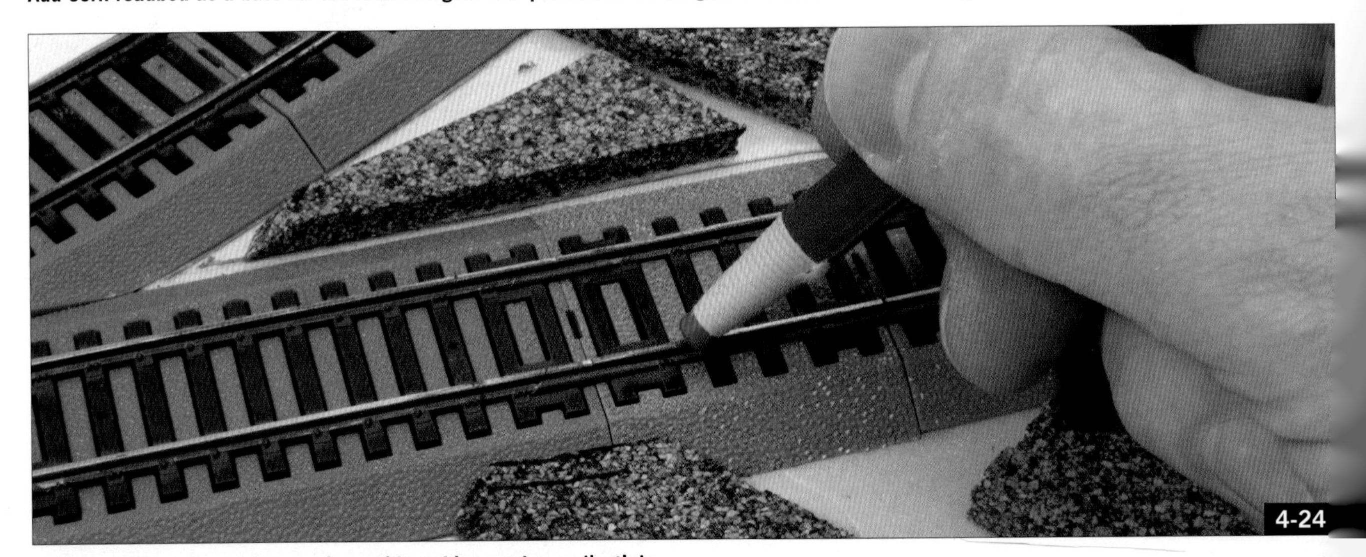

Coat the rails at the grade crossings with a china marker or lipstick.

The resulting marks on the underside serve as a pattern for cutting, **4-25**. Score the styrene at the proper marks with a hobby knife and snap it clean.

Run beads of thick CA along the cork roadbed and press the road into place. Add weights if necessary to hold the styrene securely until the glue takes effect.

I finished the grade crossings using the same material to simulate a paved crossing. The road itself should butt firmly against the outer half of the rail. Cut pieces of styrene to fit between the rails at each crossing. For a better appearance, I used two pieces laminated together to increase the height, **4-26**. Make sure the inner pieces clear the flangeways – run a freight car through the crossing to make sure.

It's a good idea to keep the top of the road slightly below the height of the rail tops. This will make it much easier to clean the track without accidentally scraping paint off the road and also lessen the chance of an uncoupling pin snagging the crossing piece that sits between the rails.

Painted highways

Paint the roads to match the concrete or asphalt surfaces they represent. Although I installed the concrete roads at this point, I recommend waiting and painting them after the scenery is in place.

Let's start with the concrete streets and roads. Paint the Walthers and Rix components with Polly Scale Aged Concrete, **4-27**. Use a wide, flat brush on the roads and a fine brush near the grade crossings. Real concrete varies in color, so it's okay if the results are a bit uneven. You can use other colors as well – the key is to use a flat color so that the final surface has no sheen.

Follow this with any necessary trim painting. With a fine brush, I painted the curbs at the intersections yellow, a common practice to indicate no-parking zones, **4-28**.

Then add the center striping. I used white stripes, since this layout is set in the mid-1960s. Stripe length and use varied by state and region, with white stripes used on two-way roads into the early 1970s and yellow

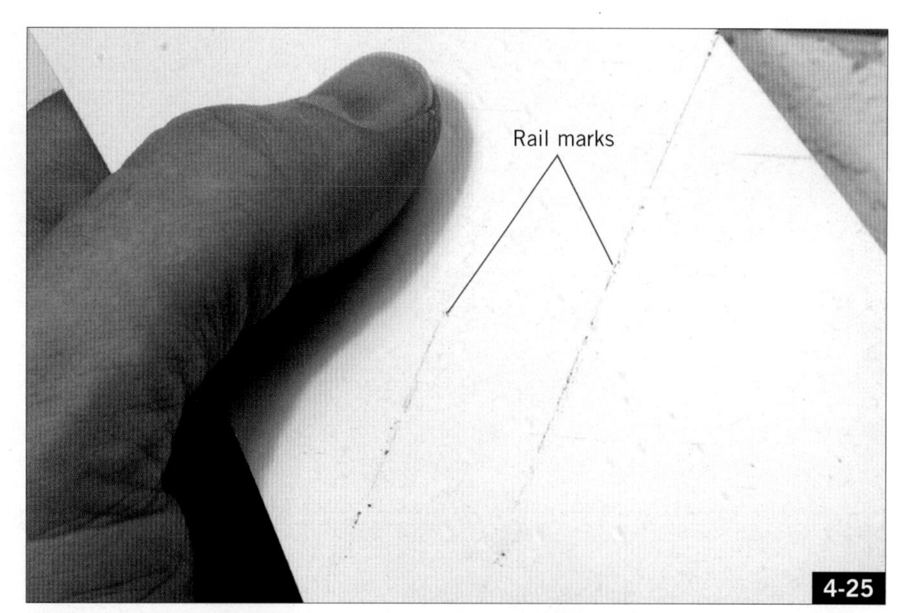

Press the road in place and the marks will transfer to the styrene to provide a cutting guide.

Cut styrene to fit between the rails to simulate paving.

A wide, flat brush works well for painting the streets a concrete color.

You can paint the curb areas yellow for no-parking zones.

Slice the dry transfer stripes to length on their backing sheet.

Rub the backing sheet with a pencil to burnish the stripe in place.

thereafter, with solid yellow lines to indicate no-passing zones.

You can use a few different techniques to apply stripes. If you have a steady hand, you can paint them freehand with a fine brush. You can also outline each stripe with masking tape to create a guide for your brush.

As I did, you can apply these markings with dry transfer stripes from Woodland Scenics, **4-29**. I carefully cut the white stripes (without cutting the backing sheet) with a hobby knife into pieces about a scale 10 feet long. You could also cut the entire sheet, which might make the stripes a bit easier to apply. Place the transfer atop the road and burnish the stripe section into place by rubbing it firmly with a burnishing tool or pencil, **4-30**. When you remove the backing sheet, the transfer will remain on the road. You can add solid yellow striping in the same manner.

I wanted my streets to appear well-used, so I used other concrete colors (such as Polly Scale concrete and concrete mixed with a bit of white) on some road sections to make it look like some slabs had been removed and replaced. A fine brush and some grimy black paint simulates cracks and expansion joints that have been tarred, **4-31**. I also drew some cracks in the surface with a pencil, **4-32**. Vary their number and extent to match the type and age of road you're creating.

The next step is to weather the road by simulating the marks left by tires, oil drips and spills, and general deterioration. I used black, gray, and white powdered chalk for this. I placed some on a piece of cardboard and blended them to make various shades.

Take a standard two-inch angled paintbrush, dip it into some chalk, and gently touch it to the road, **4-33**. Use a quick brushing motion along the road to make the wear marks found along the highway lanes. Using a large brush like this is faster and offers more control than a smaller brush. Repeat the process until the effect is what you're looking for, **4-34**. If the appearance is too dark or doesn't turn out quite right, just remove the chalk with a damp cloth and start over.

The painting process is similar for an asphalt road. Asphalt varies greatly in color depending on its age and composition. And, like concrete, no asphalt road is a single color – look closely at a

Add tar to the expansion joints and other locations with a fine brush and grimy black paint.

Pencil lines simulate cracks in the concrete that haven't been tarred or repaired.

A large brush and some powdered chalk create the lane wear marks often found on streets.

The finished street and road look like they've seen a few years of service.

A small foam roller does a great job of evenly painting the highway surface.

Serrated knives work well for cutting foam. Find a knife with a heavy, stiff blade.

Stanley Surform tools shave off layers of foam, making them ideal tools for final shaping and contouring.

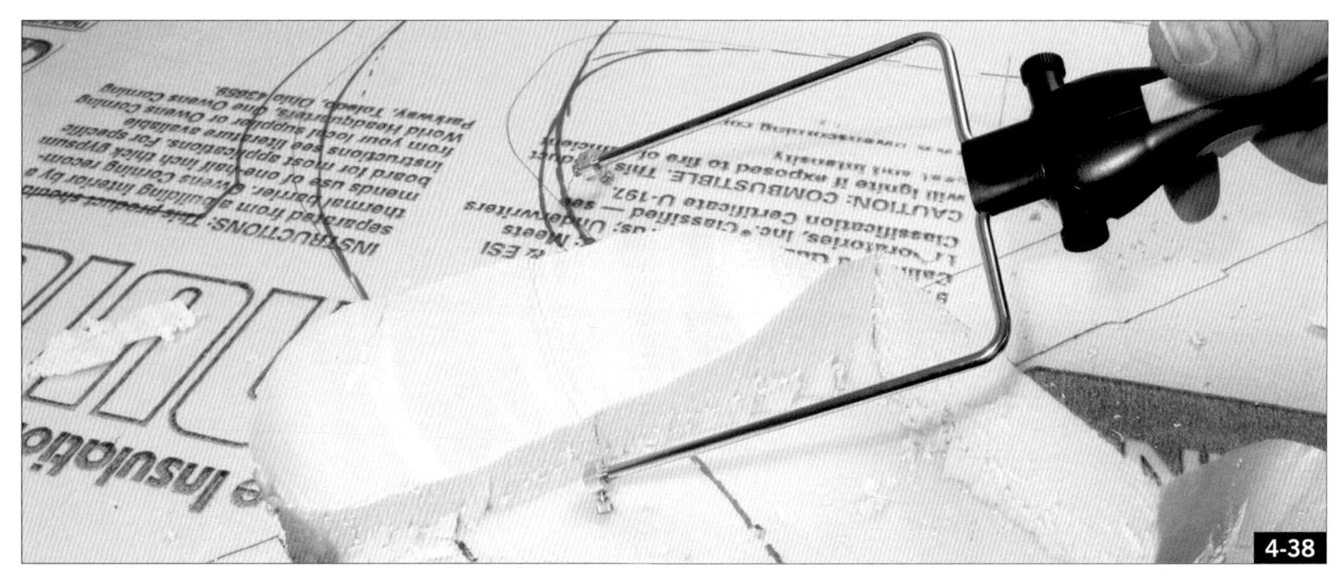

Note how clean the cut surface is from this Woodland Scenics hot wire tool.

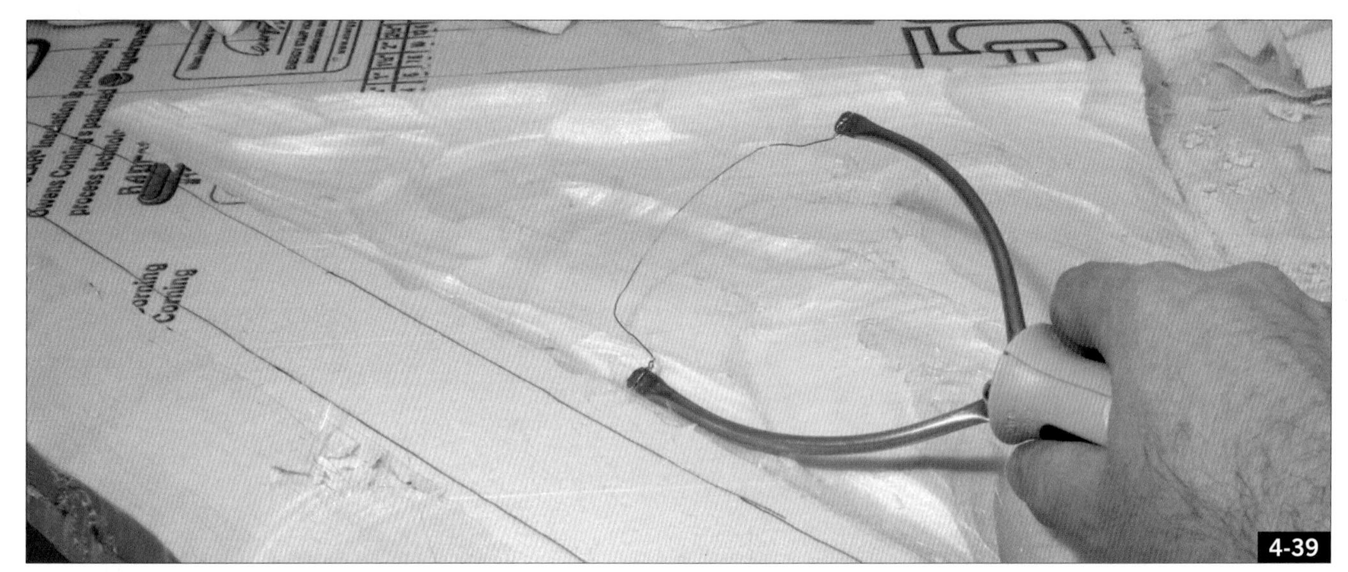

The wire on the Hot Wire Foam Factory router tool is easily shaped, making it simple to cut these ditches and recessed areas.

The foam is contoured to the edge of the layout, as shown here where the road will pass under a railroad bridge. Note the ditches along the highway and railroad embankment.

4-41 To make hills, simply cut and shape pieces of foam and then glue them to the table. Note the ditch cut between the track outline and the new hill.

real road and you'll see that the aggregate making up the surface is actually dozens of different colors.

You can mix model paint, such as Polly Scale white and grimy black, or interior house paint to the shade desired. I decided to mix flat white and black household latex paints that I had on hand. The key with either paint is to use a flat finish. Apply the paint with a small, inexpensive foam roller (available at craft stores) to give it a uniform appearance, **4-35**. It took two coats to completely cover the styrene.

Weather the asphalt road in the same manner as the concrete streets, using various shades of chalk to give the road a mottled look. Apply the stripes in the same manner as well.

Contouring foam

As you begin carving foam to shape the layout's foam scenery base, it's wise to have a clear picture in your mind of the final shape and scenic contour that you're looking for. If you're trying to re-create a scene from real life, you can follow photos of the actual scene. You can use another model railroad as inspiration or simply create a scene that you picture in your mind.

4-42 Test-fit the abutment, making sure the track passes over its top. You'll also need to adjust its overall height.

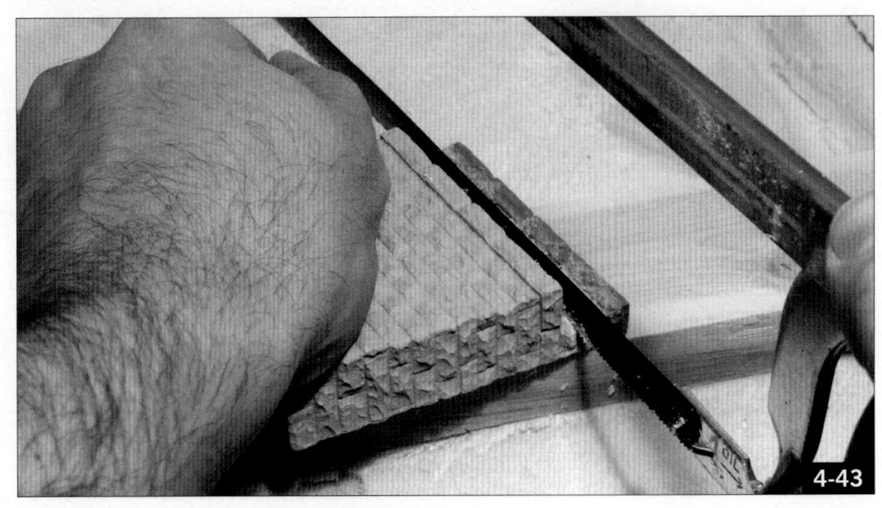

4-43 If needed, trim the top of the abutment with a hacksaw.

A number of tools work well for carving foam. A steak knife with a serrated edge cuts foam quickly, and it's good for major surgery, **4-36**. For final shaping and contouring, you can use Stanley Surform tools, which are available in a variety of sizes, **4-37**. The Surform tools have a cutting surface similar to a cheese grater for clearing the cut material. The downside with both these tools is that they generate many small pieces of foam that seem to get everywhere. Keep a shop vacuum handy and use it frequently to keep from tracking foam dust throughout your house.

For neat cuts in foam, it's hard to beat a hot wire foam cutter. These tools send electricity through a thin wire that heats up to melt the foam and cut it cleanly. For most cutting, a basic light-duty tool is all that is needed, **4-38**.

For contouring and shaping foam below the table's surface, or for making deep cuts in foam, it's handy to have a tool with heavy wire that can be bent into several shapes. I made good use of the Hot Wire Foam Factory's router tool, **4-39**. Being able to shape the wire made it easy to cut recessed ditches, cut the highway grade, and contour the railroad embankment on the layout's right side.

When cutting extruded polystyrene, such as pink and blue insulation board, with hot wire tools, be aware that the smoke and fumes generated are harmful to breathe. Only use these tools with adequate ventilation. It's also a good idea to wear a cartridge respirator while carving foam.

There is no magic formula to contouring foam – just keep removing foam until you have the desired shape, **4-40**. Once you cut the basic shape with a hot-wire tool or knife, use the Surform tools for final smoothing and shaping.

Make sure the bridge sits level atop the abutments but don't glue the bridge in place until the scenery is completed.

The grade separation of the finished scene adds to the layout's realism.

Temporarily clamp the fascia in place and trace the outline of the scenery onto it. The foam block cushions the clamp that's holding the fascia.

Bevel the corners of each fascia panel at a 45-degree angle for a clean, tight-fitting corner.

Cork works well as a base for structures and parking lots. Mark the structure location before adding scenery.

Removing too much foam isn't a problem, as you can always glue more foam in place and start again. The contour doesn't have to be perfect, as it will be covered with ground material in a later step.

You can easily add hills above the table level by contouring a piece of foam, cutting it to size, and gluing it in place on the table. I did this in one corner of the layout, **4-41**. Note the ditch – cut with a hot-wire tool – between the hill and the track outline.

Trim spare pieces of sidewalk material as needed for extra steps, such as these on the corner bank building.

Adding a bridge

Variations in elevation add to realism, and one great way to achieve this is by adding a railroad bridge over a road or waterway. I decided to add a bridge over the road on the right side of the layout.

The first step is to select a bridge. Railroads generally used through bridges over roads and other railroads to maximize clearance under the bridge. I decided to use a Central Valley through plate-girder bridge, an injection-molded styrene kit (no. 1903) that I'd assembled for an earlier project. If you don't want the challenge of assembling a kit, you can make the project a bit easier by using an assembled through plate-girder bridge, such as an Atlas no. 592.

Once you have the bridge, determine where you want it. If you're using an Atlas bridge, it's easy – the bridge has the track built in, so it simply replaces one of the 9" track sections. The Central

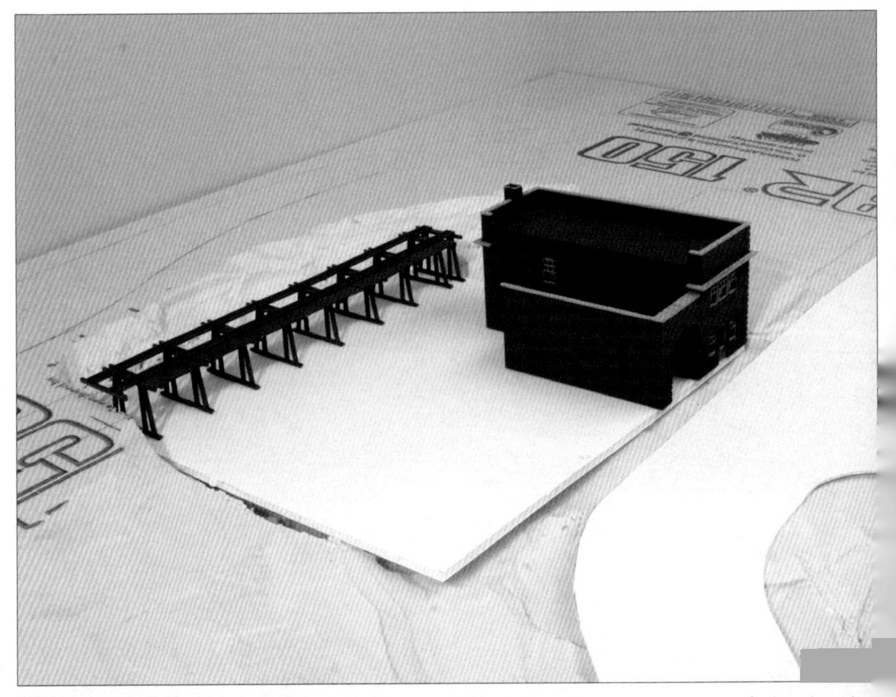

The coal yard and main building rest on a foam core base. Make sure the base is level and that the trestle matches the track height.

Valley bridge doesn't include track, so the Atlas track (with the roadbed cut away) will pass across the bridge deck. This means the bridge can be located anywhere along straight track. Prior to cutting the foam, I placed the bridge over the track (which wasn't yet glued in place) and marked its ends on the foam. You'll also need to allow for the thickness of the top of the abutments.

The bridge ends need to rest on a pair of abutments. I used a pair of Chooch cut-stone abutments (no. 8440). Test-fit the bridge on the abutment with a piece of track and roadbed on the backside of the abutment, **4-42**. The track should pass over the top of the abutment and rest on the bridge deck while the bridge shoes rest firmly on the abutment.

I had to cut one course of stones out of the back retaining wall to get the "step" on the top of the abutment to match the height of the Central Valley bridge. A regular hacksaw works well for this, **4-43**. Follow the separation and then use CA to glue the top back in place. You'll probably have to also cut the abutment walls down to the proper height – this will vary depending upon how much foam you've carved away below the bridge.

Test-fit the bridge atop the abutments, **4-44**, and when everything is properly aligned, glue the abutments into place on the foam. Make sure they're vertical and that the flat platform (upon which the bridge shoes rest) is level. You can then cut the track roadbed to fit against the outside walls of the abutments. You can lay the track up to the abutments, but don't add the bridge until you have completed all of the surrounding scenery, **4-45**.

Fascia

Once the foam is contoured, it's time to apply fascia panels around the edge of the layout. The fascia gives the layout an attractive, finished appearance by hiding the benchwork frame and edges of the foam. By adding the fascia now, before beginning the scenery, you can easily blend the scenery to the edge of the layout.

I used the same foam core for the fascia as for the backdrop because it's lightweight, simple to cut, and easy to

Secure the foam core base to the layout with Foam Tack Glue.

Cut one panel from a City Classics retaining wall to use as an abutment.

Glue the abutment in place so that it supports the end of the coal trestle at the proper height.

4-54

The Woodland Scenics corner barbershop is glued to a foam core base with scenery materials to represent a neighboring vacant lot.

glue. As with backdrops, you can also use other types of flat sheet material for fascia panels, such as hardboard, thin extruded foam, plywood, or thick styrene sheet.

I began by cutting the foam core into strips 6¼" wide. This is deep enough to cover both the foam and the 1 x 3 frame and allows for a slight overhang at the bottom that hides the tops of the bookcases. If you only used one layer of foam, you can make this shallower (about 5").

Cut a piece to length to cover each straight segment on the front of the layout. Clamp each piece into position, and mark the back side with the profile of the foam along the edge, **4-46**. Remove the piece and use a sharp hobby knife to cut the fascia to match the profile of the foam.

For the best fit, join corner pieces with beveled edges. Cut the outside corners with a bevel, simply angling a hobby knife at a 45-degree angle against a straightedge while cutting the corner, **4-47**. Fasten each fascia panel by spreading Foam Tack Glue along the 1 x 3 and the foam edges and pressing the panel into place. Lightly clamp the panel if necessary, being careful not to dent or mar the foam core.

Any gaps at the joints between panels can be filled with Sculptamold or Woodland Scenics Foam Putty when adding this material as the basic scenery coat.

Building foundations

Structures require a solid and level foundation. If you're working with a level layout, cork roadbed and sheets work well for raising structures to street or track level. And if needed, you can use multiple layers of cork to raise a foundation to the proper height. Once a structure is in place, outline the building on its base with a marker, **4-48**. You'll actually add the building when the scenery is in place to minimize the risk of damaging the structure during the scenery process. I used the following method for the factory structures on each end of the layout.

Since the storefront buildings rest on the same cork base as the street, all I had to do was shim them to the level of the sidewalk. A couple of them had bases that could be trimmed (see pages 38-39). For buildings without bases, I simply used leftover sidewalk pieces from the Walthers street kits. For example, the corner bank building needed some small sidewalk pieces cut to match the building's steps to the Walthers sidewalk, **4-49**.

Some structure installations are trickier. The coal yard is one example, as the building sits on a hill, and the coal yard is lower than the track spur that enters the yard on a trestle. To fit, I measured the room needed for the trestle. Figuring in the height of the track and roadbed, the surface of the coal yard needed to be 1¼" below the original top of the foam table.

Cut a piece of thick foam core as a base for the coal yard and test-fit it at the location, **4-50**. Then trim the foam table until the height is correct. You can use scraps of foam or foam core to shim the base to the proper height, making sure that it's level. Glue the base into place using Foam Tack Glue, **4-51**.

Add a retaining wall where the coal trestle butts against the foam, so the foam doesn't show. I used a City Classics injection-molded retaining wall (no. 601) cut to one panel wide, **4-52**. You'll also need to adjust the height and then glue the panel in place. Make sure the top of the trestle stringers are at the same height as the bottom of the ties on the adjoining track, **4-53**. Set the trestle aside for now.

I used a similar technique for the car dealer on the hill, cutting a piece of foam core to match the size of the building base. This base also required some shims before being glued in place on the contoured foam.

The hardware store and barbershop on the hill along the street were also a challenge. For these, I opted to mount each building on a base before installing them on the layout. For the corner barber shop, I used a similar technique as the hardware store but added a vacant lot on one side with some scenery added to it, **4-54**. This can be handy for structures in tight areas or for those located deep into a layout.

Adding these buildings on the hill is tricky because they must remain level even though the road is on a grade. As a result, part of the foam core base on each building will be visible. To make them look like foundations, I painted them to represent concrete.

Carve the foam subbase under each building to allow it to sit as level as possible, **4-55**. You can also add foam or foam core shims to keep things level. Once the fit is good, glue the shims in place, **4-56**. However, wait until you've finished the scenery behind the structures before gluing them in place.

In the next chapter, we'll complete the scenery by adding texture and some finishing touches.

If necessary, use shims to level the buildings on the hill. Note the Walthers sidewalk sections added to this part of the asphalt road.

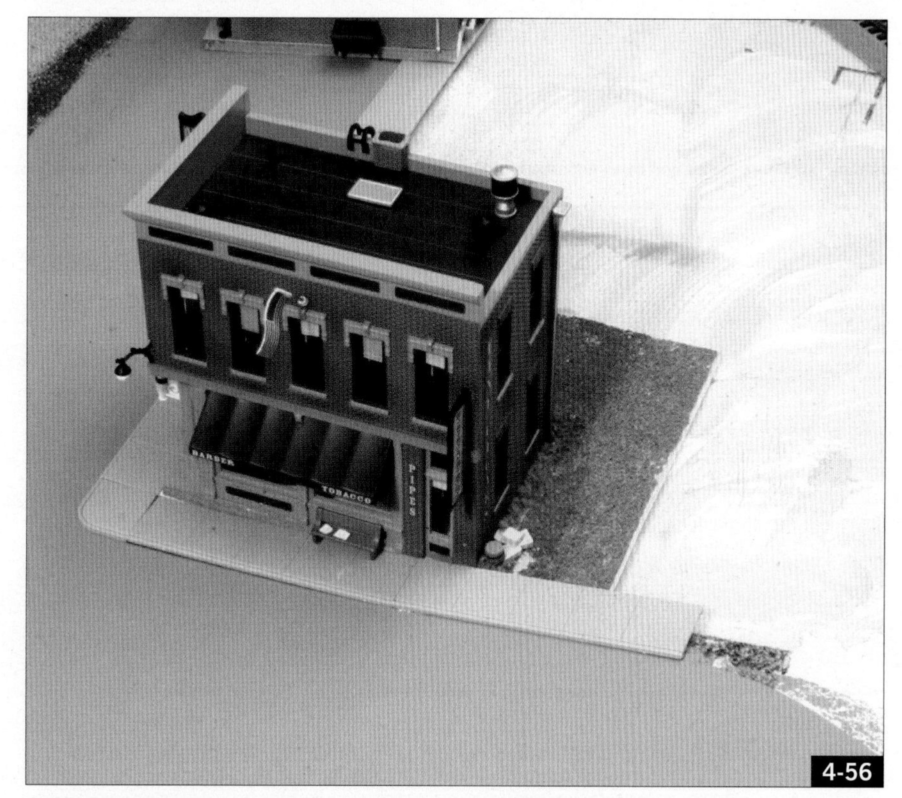

Test-fit each structure to make sure it sits level and tight against the sidewalk.

5-1

CHAPTER FIVE

Scenery texture and final touches

Texture and color are both important in setting a scene. The variation among the farm field dirt, track ballast, grassy area, concrete roadways, and gravel parking lot all contribute to the realism of this scenery.

Once the basic layout shape is in place, the next step is to finish the surface with colors and textures that match the look of grass, weeds, brush, dirt, and other real scenery, 5-1. Several manufacturers make a wide variety of scenic products for doing this. Popular materials include ground foam and static grass in various colors and textures, cast rocks, and the ever-popular (and free!) real dirt and gravel. Let's start by cleaning up the layout surface in preparation for texturing.

5-2

Sculptamold is a powdery, fibrous material. To use, mix it with water in a small plastic bowl.

5-3

Add water slowly until it has a thick, paste-like consistency. It shouldn't be runny.

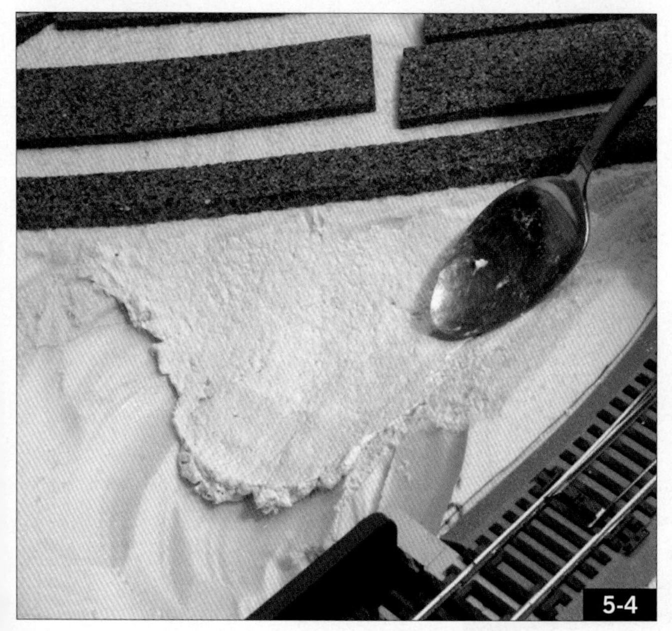
5-4

Use a spoon or spatula to spread the Sculptamold on the layout surface.

5-5

Sculptamold does a good job of smoothing areas such as the joint between the layout and abutments and the edges of the road.

Scenery surface

The layout is almost ready for texturing with scenery, but first, we must provide a smooth base for grass and other scenic elements. In many areas on the table, the foam is suitable for a scenic base. However, other areas might have jagged edges. There may also be gaps around bridge abutments or sharp angles around structure bases and road and street edges. These areas need to be filled in to create a level base.

Sculptamold is my favorite material for a scenery base. This is a fibrous material that you add water to, mix, and spread on the surface, **5-2**. Although it's a bit more expensive than plaster, Sculptamold has several advantages. The biggest one is that it's not as messy. When mixed, it holds together well, and won't drip like plaster, so there's no need to cover the layout's track or any carpet on the floor. It's also lighter and easier to shape and contour.

Start by mixing a small batch in a plastic bowl. Slowly add water to the Sculptamold, mixing until it has a thick, pasty consistency, **5-3**. If the material is soupy, drips, or runs, you've used too much water.

Apply the material where needed on the surface with an old spoon, **5-4**. The back of the spoon works well for contouring and shaping. You can also use a spatula, putty knife, or even your fingers to form the material.

Keep adding the Sculptamold in small batches – about as much as you can apply in 10 minutes or so, **5-5**. Use it to fill any gaps between the foam surface and the fascia or even areas around structure bases. You can also add it randomly atop the flat foam surface to provide some variance in contour.

From left: Woodland Scenics fine turf in a shaker bottle and coarse turf in a bag, Bachmann blended turf in a shaker bottle, Noch static grass, and Scenic Express turf in an applicator bottle.

Paint the ground with a thick coat of flat tan paint, working it to the edge of the roadbed and to structure outlines.

Basic scenery texture

After the Sculptamold dries, you're ready to make the layout come to life by adding a coat of green texture. Woodland Scenics, Bachmann, Scenic Express, Timberline, and others make a wide variety of ground foam materials in various shades, **5-6**. Start with fine and coarse ground foam in several shades, depending upon the area and season you want to represent. To represent summer, I used Woodland Scenics green, light green, green blend, weeds, and soil.

The first step is painting a small area (about a square foot or so) with earth-color flat latex paint, **5-7**. Don't waste money on the good stuff – buy the cheapest flat tan paint you can find, as you'll be using it like tinted glue. Just make sure that it's water-based and dries with a flat finish.

Apply the paint liberally with a foam brush, working it into all the nooks and crannies in the surface. Foam brushes are inexpensive, clean up more easily and quickly than standard paintbrushes, and do a good job of spreading paint on rough, uneven surfaces.

Sprinkle ground cover over the wet paint, beginning with fine foam. Blend colors for a varied look. A Parmesan cheese container makes an excellent applicator.

Begin sprinkling the ground texture in place while the paint is still wet, **5-8**. Start with fine ground foam of the desired color. I like to use at least two shades to get some variation in the color, much as you see in real life – in this area, I used green blend and weeds (a darker green color).

Woodland Scenics and others make large plastic applicator bottles with holes in the top for sprinking the foam. I use these as well as old Parmesan cheese containers that work beautifully (and are cheap!). Along with ground foam, you can use fine ballast to represent gravel in parking areas, **5-9**. A small cup works well for this.

Depending upon the texture you're looking for (overgrown ditch, weedy lot, neatly trimmed yard), you can add some coarse ground foam as well, but it's often easier to add this after the glue is applied. Coarse and extra-coarse foam are good for representing deep weeds, shrubs, and scrub brush.

Spray the ground foam with a wetting agent. The standby for years has been what many modelers call "wet water," which is tap water with a few drops of

Various colors of fine ballast, or real dirt and gravel, can represent gravel roads and parking lots.

A pump-type spray bottle that delivers a fine mist is ideal for wetting the ground cover.

dish detergent added. (The detergent breaks up the surface tension of the water, allowing it to readily soak into the scenery materials.) You can use this, but I've had much better results using full-strength rubbing alcohol. I've found that the alcohol doesn't disturb the scenery materials as much as the water mix.

Use a pump-type sprayer that delivers an extremely fine mist. I've had the best results with an old hair spray bottle, **5-10**. Most of the commercial trigger-type sprayers I've used don't atomize the spray fine enough, and the larger droplets they emit can disturb the scenery as they land.

Keep the sprayer well above the scenery, and use gentle "poofs" to avoid blowing the foam out of position. Mist the ground foam until it's wet. You can use a scrap piece of cardboard or plastic to keep the spray off the backdrop, structures, and other areas, **5-11**.

Next comes a coat of adhesive to lock the ground foam firmly in place. I use white glue mixed with water (about three parts water to one part glue). You can also apply thinned acrylic matte medium (available in art supply stores) or Woodland Scenics Scenic Cement.

A scrap piece of plastic or cardstock keeps spray off of the backdrop.

An old saline eyewash bottle works better than an eyedropper for applying diluted glue to the wet ground cover.

Dribble the glue onto the wet ground foam. An empty spray bottle for saline eyewash or contact lens solution works very well for this, **5-12**. The tip is perfect for applying drops, and you don't have to constantly pause to refill it as you would if you used an eyedropper.

Make sure the area is thoroughly soaked by the glue mixture. It will look like a mess when initially applied, but as the glue dries, it disappears, leaving the scenery texture looking good.

While the area is still soaked with glue, add additional fine or coarse foam. Coarse and extra-course foam can be pressed into place, and you can add more wetting agent or glue as necessary.

Static grass presents a much better appearance of individual blades of grass than does ground foam. This is made of small green fibers that, when applied with an electrostatic applicator, stands up on the scenery surface. Apply static grass over the ground foam while it's still wet with glue, **5-13**. Several brands

Apply the static grass by lightly squeezing the applicator bottle above the glued scenery area.

Static grass gives the appearance of individual blades of grass, while the coarse ground foam in the background looks more like brush.

are available; I used Noch, which comes in plastic applicator bottles that can be refilled from bags of the company's grass fibers.

Fill the applicator bottle with the static grass material. Hold the bottle five or six inches above the ground surface and lightly squeeze it, "poofing" the grass into place. The glue holds it, and the static electricity of the plastic bottle pulls the fibers upright. You can see the final effects of this in photo **5-14**.

Repeat the process in small patches until the entire area is covered. Extend the scenery to the lines marking the structure outlines but keep the scenery bases wiped clean of spray, glue, and texture materials. You can touch up these areas after the buildings are in place.

Plowed field

I wanted to add a plowed field in one corner to give the layout a typical Midwestern scenic element. To replicate the look of real dirt, I used...real dirt, **5-15**. The outdoors is a great (and free) source of many materials, especially soil, sand, rocks, and various types of gravel.

Rub the dirt through an old piece of window screen to sift out the largest chunks. Then apply it to the field in the same way as the other ground scenery. Sprinkle the dirt atop the paint and pat it down with your hand to press it into the paint.

To give the appearance of plow furrows, take a comb, remove every other tooth, and scrape it across the field, **5-16**. Repeat the process where necessary and spray the field with wetting agent and bond it with glue.

If you want to add the appearance of soybeans or other crops, apply lines of Woodland Scenics Scenic Cement across the dirt and sprinkle coarse green foam on it.

Ballast

The all-in-one track and roadbed has molded-in texture, so it's not necessary to ballast the track.

In real life, industrial spurs often look different than main lines. They use different ballast, and they are often overgrown with dirt and grass up to (and sometimes over) the ties.

5-15

Real dirt can be used as scenery. Sift it through a fine screen to remove rocks and oversize chunks.

5-16

A comb dragged across the surface gives the appearance of a plowed field.

To give the industrial spurs a different look on the layout, I poured Woodland Scenics fine cinders ballast (black) into a small plastic cup until it was about two-thirds full. I added fine ground foam (weeds) until the cup was almost full, covered the top with my hand, and shook it until it was thoroughly mixed, **5-17**.

Apply this grassy cinders mix directly over the track. Use a soft brush to clear the ballast off the rails but keep some on the ties. Bond the ballast as with the scenery, spraying it with alcohol and adding diluted glue, **5-18**. Roll a freight car across the track to make sure no stray ballast interferes with operation and remove any that does.

Adding structures

Once the basic scenery is completed, the structures can be added. Although buildings can simply be set in place without bonding them, their appearance often won't be realistic, as gaps may show between the structure base and the scenery. Buildings may also shift or move if they're not glued in place.

5-17

A mix of fine cinders ballast and fine ground foam is ready for applying to a spur track.

5-18

The finished track has the look of an overgrown, little-used spur. You can vary the effect as desired.

Add glue along the structure base and place ground foam or ballast in the glue.

The ground foam hides any gaps between the structure and layout.

First, test-fit each structure in its location. Add a bit of Scenic Cement to the base, or bottom of the walls, and press the building into place. You don't need much glue – just enough to keep it from moving.

To blend a building to the adjoining scenery, hide any gaps around the structure's base with ground foam or ballast. For grassy areas, add some Scenic Cement to the base and press coarse ground foam into the glue, **5-19**. Any gaps will be hidden when the glue dries, **5-20**.

If you need to add gravel, ballast, or fine ground foam around a structure or next to a road, first add the material and then apply rubbing alcohol with an eyedropper, **5-21**. Follow this with an application of thinned white glue, **5-22**. This method gives you more control than when spraying the alcohol.

Prior to adding the trestle at the coal yard, I added a very thin layer of fine ground foam (soil) to the area under the trestle since that would be tough to do after the trestle was in place. I bonded it as with the other

After adding ballast or ground cover to existing scenery, apply alcohol to the area with an eye dropper or pipette.

Follow the alcohol with a coat of diluted glue to hold the materials in place.

Weight the trestle down until the glue sets and the ground foam is dry.

Add coal piles beneath the trestle by pouring scale coal between the stringers.

scenery. I then spread some Scenic Cement on the bottom of each trestle bent, placed the trestle in position, and put weights atop it to hold it in place until the glue dried, **5-23**.

Once that dries, you can add coal piles by pouring Woodland Scenics scale coal through the trestle with a small cup between several bents, **5-24**. Wet this with alcohol from a spray bottle or eyedropper and then use an eyedropper to carefully add diluted glue to the piles, **5-25**.

Add the track atop the trestle. Run a bead of CA along the stringers of the trestle, set the track in place, and add some weights to hold it until the glue dries. While you're at it, you can add the track across the rail bridge as well and glue the bridge in place with small drops of CA under each bridge shoe.

After the trestle, track, and coal piles are dry, you can add final details to complete the coal yard, **5-26**.

Trees

The availability of good-looking, ready-made trees has made it pos-

5-25

Bond the coal by carefully applying alcohol and then diluted glue to each pile with an eyedropper.

5-26

The coal piles under the trestle, along with the extra details from the Walthers kit, bring the scene to life.

5-27

Ready-made trees from left: Woodland Scenics standard and premium trees and a Bachmann sycamore.

5-28

Poke a hole in the scenery where you want to plant a tree.

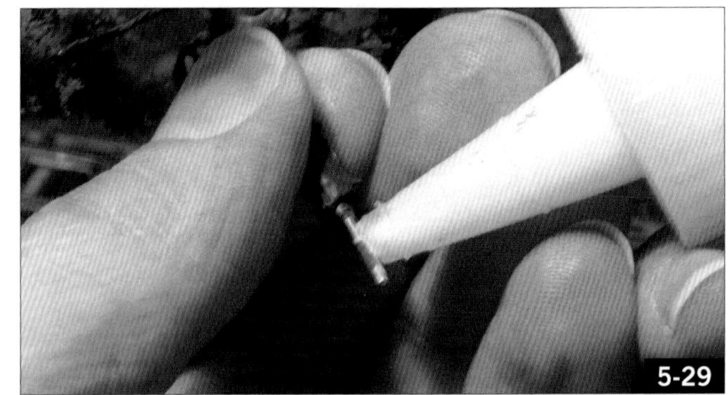

5-29

Add a bit of glue to tte tree's mounting wire or peg and stick it in place on the layout.

5-30

Ground foam will hide any gap at the base of the tree.

sible to complete scenery much quicker than in the past. Assembled trees are available from Bachmann, Busch, Faller, Heki, Noch, Timberline, Woodland Scenics, and others. Some are generic in appearance, but many are modeled after specific varieties of coniferous or deciduous trees, **5-27**.

Most of these trees have plastic or metal mounting pegs at the base of their trunks, which in many cases, are designed to snap into plastic bases. The best way to install these trees is to discard the bases and mount them directly to the layout. Poke a hole into the foam with an awl or other sharp pointed tool, **5-28**, place some Scenic Cement on the mounting peg, **5-29**, and push the tree into place. If there's a gap at the base, add a bit of glue and some additional ground foam to hide it, **5-30**.

I like to place the generic (and less expensive) trees in the background and in groups and use the nicer premium-quality trees in the foreground and by themselves.

Trees can help hide the joint between the backdrop and the layout. Gluing pieces of foliage against the backdrop keeps viewers from seeing blue sky among the trees, **5-31**. By planting several trees in front of the foliage, you'll give the appearance that there are actually more trees there, **5-32**.

5-31

Glue Woodland Scenics foliage against the backdrop in areas you want to forest.

Final touchups

Once the basic scenery is complete, look for any areas that need touching up. Take a bottle of Scenic Glue and a pack of green coarse ground foam and go around the layout.

The backdrop is a good place to start. Run a line of glue on any areas where the layout and backdrop joint can be seen or where stray paint or Sculptamold has gotten on the base of the backdrop, **5-33**. Press the ground foam in place against it, **5-34**. You can make this line of foam quite high on the backdrop since, in the background, it tends to fool the eye into thinking that the scenery continues into the backdrop, **5-35**. Also examine the backdrop for any stray marks or glue stains and touch them up with some blue paint.

Check the areas next to roadways and sidewalks, **5-36**, and any areas where the ground meets bridge abutments and retaining walls. Add ballast or dirt to parking lots and gravel roads as needed.

5-32

The foliage helps hide the backdrop from view when looking among the trees.

Spread Scenic Glue along the backdrop where it meets the scenery.

Once the scenery is done, make sure everything still runs fine. Clean the track thoroughly with a Bright Boy or other track cleaner, **5-37**, and run a train around the layout to make sure that none of the scenery or structures interfere with operations.

Then paint the fascia. I used flat black interior latex, applied with a small foam roller. I think that black works well to frame the layout, and it draws viewers' eyes to the scenery above it. Other modelers paint the fascia dark green to match the basic scenery color or dark tan or brown to match the basic earth color. Try a color, and if you don't like the effect, just repaint it with a different color.

We're getting close. Let's look at locomotives and rolling stock.

Press coarse ground foam into the glue with your fingers or a tweezers.

The ground foam helps hide the joint between the layout surface and backdrop.

Touch up any areas as needed, such as along the roadbed or highway shoulders.

When the basic scenery is complete, clean the track with a Bright Boy or other track cleaner.

6-1

CHAPTER SIX

Locomotives and rolling stock

Manufacturers have produced a wide variety of nicely detailed, smooth-running diesel and steam locomotives. (See the key on opposite page to identify locomotives.)

Locomotives along with freight and passenger cars have led the way in the growth in number and quality of ready-to-run models. Into the 1990s, getting a well-detailed locomotive meant adding lots of aftermarket details; for freight cars, it meant building a kit and applying additional details. That situation has changed dramatically, as most manufacturers now offer products assembled with a high level of detailing.

Locomotives

Today's high-end ready-to-run locomotives feature a plethora of separately applied details that include horns, antennas, bells, grab irons, lift rings, door levers, windshield wipers, and cab interiors with figures. Shell detail and molding quality have increased markedly, with such items as separate fan blades, etched-metal grilles, and flush-fitting window glazing. Many models include optional or prototype-specific details, such as multiple types and styles of headlights, horns, cabs or cab windows, pilots, different sizes of fuel tanks, and different styles of trucks. These high-quality models are available from Athearn Genesis, Atlas, InterMountain, Kato, Proto 2000, and Stewart, **6-1**.

Ready-to-run steam locomotives have also become well-detailed models. Many feature extensive piping, bell ropes, valve gear, compressors, bells, whistles, and blackened metal rods and wheels. Many include detailed cab interiors with flush-fitting window glazing and crew figures. As with diesels, some steam models are patterned after specific railroads' locomotives, with representative pilots, lights, wheels, cab styles, and valve gear. Multiple tenders are often available.

Although they don't include as much detail, many mid-priced locomotives still run very well. The Athearn Ready-To-Roll line, Atlas Trainman series, Bachmann Spectrum, International Hobby Corp. steam locomotives, and Proto 1000 locomotives all fit this category. Most of these models feature accurate paint schemes, and they are excellent starting models for those who enjoy detailing and upgrading.

Deciding what locomotives to run is largely a matter of personal preference. Many modelers, especially those who've been in the hobby for a while, stick to buying only models of locomotives actually used by a certain railroad during a specific era. Other modelers aren't as restrictive, buying various engines that strike their fancy. The prime limiting factor is the radius of the curves on your layout. The layout in this book has tight curves, an 18"-radius minimum, so you need to stick to four-axle diesels

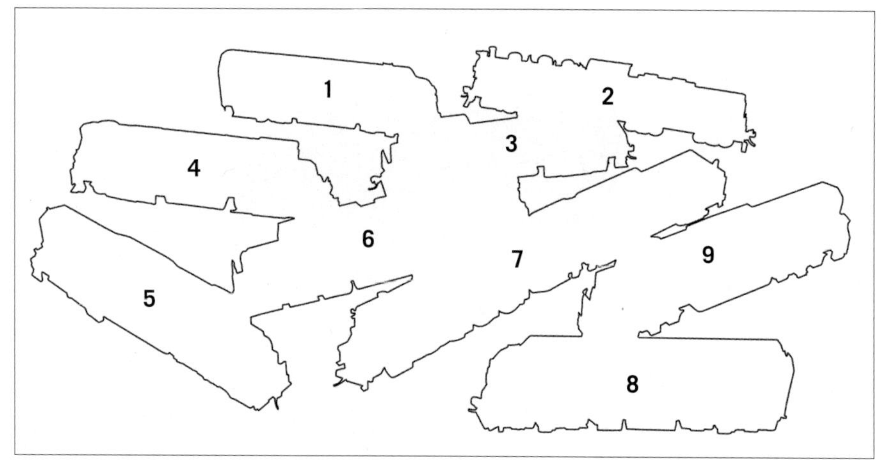

1. Athearn Genesis EMD F7, *2.* Proto Heritage USRA 0-6-0, *3.* Proto 1000 Alco RS-2, *4.* Proto 2000 Alco RS-27, *5.* Atlas EMD SD35, *6.* Atlas Master Series Silver EMD SD24, *7.* Bachmann Spectrum 2-10-4, *8.* Proto 2000 EMD F7, *9.* Athearn Ready-To-Roll EMD GP60M.

Many new locomotives include an eight-pin socket to make it easy to add a plug-equipped DCC decoder, such as this Digitrax DZ125.

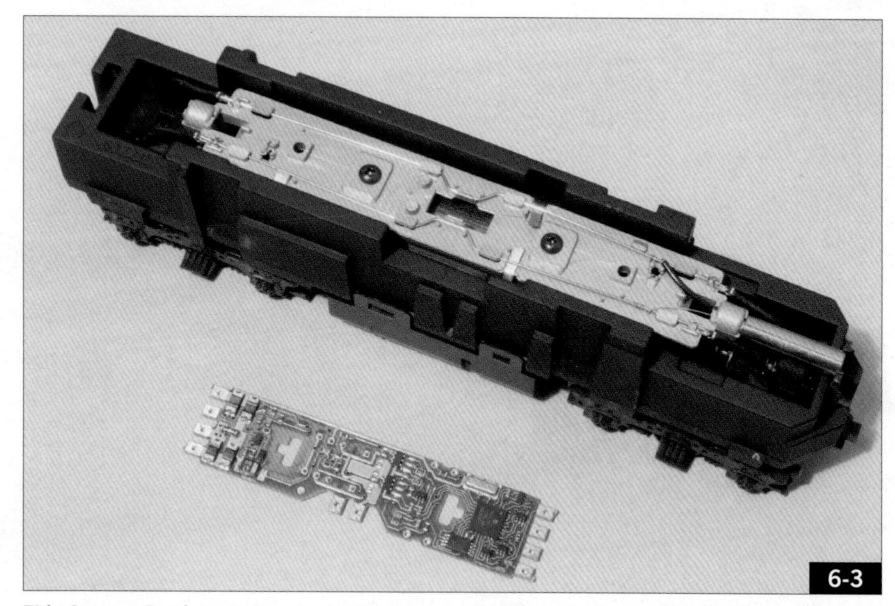

This Stewart F unit and other locomotives have wiring boards that can be replaced with circuit-board decoders, such as the Digitrax DH163A0.

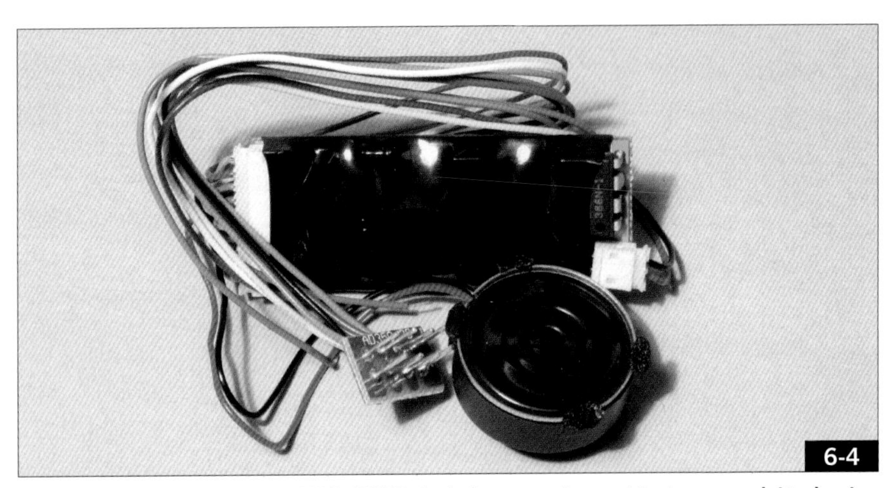

6-4 This diesel sound decoder, an MRC AD370, includes a speaker and features an eight-pin plug for easy mounting.

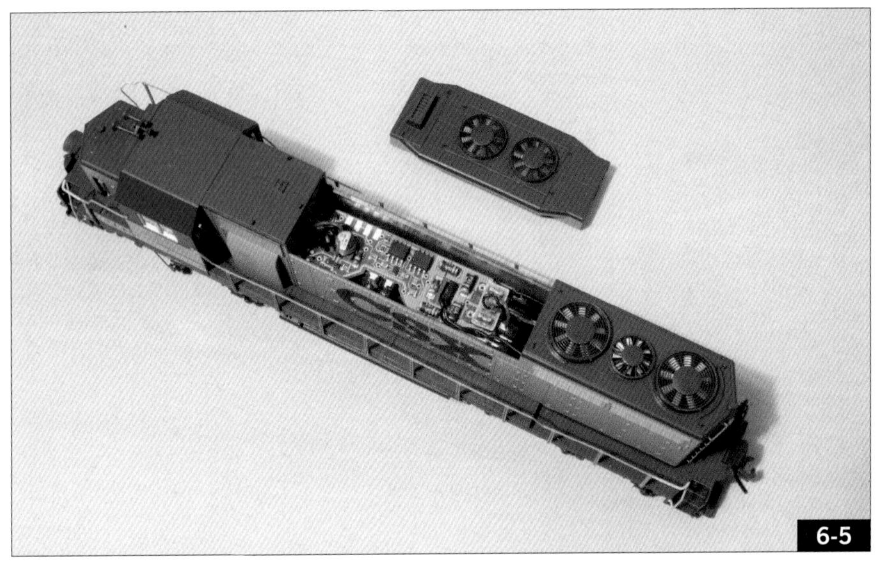

6-5 This Atlas SD35 has one of the company's factory-installed dual-mode decoders. It can be switched from standard DC to DCC operation.

and short steam locomotives (no larger than 4-6-2 or possibly 2-8-2). Larger locomotives don't look as good on the tight curves, and some may not operate reliably because of truck pivot limitations and coupler overhang.

Control

Electric and electronic advancements mean today's locomotives run much better than models of a generation ago. The can-style, or enclosed, motors that power modern models run smoother, operate at slower speeds, and draw less current than the open-frame motors of older models.

Locomotive models have paced the rapid growth of Digital Command Control. Keep in mind that even if you don't plan to use DCC on the layout you're currently building, you may want to use it in the future. Buying locomotives equipped with decoders or sockets for decoders will make it easier for you to make the transition.

Most locomotive models include a socket to simplify installation of a DCC decoder, **6-2**. For these models, you can choose almost any decoder with a matching plug (usually an eight-pin, dual-in-line plug) that has a current rating at or above the model's current draw. A 1-amp decoder is sufficient for most HO models. Check each model's manual or instructions for details, as some engines with multiple lights or other features may exceed that current draw.

6-6 Sound decoders and speakers come factory installed on many models, including this Proto 2000 EMD E6 passenger diesel.

Get access to the wheel bearings and gears by removing the cover plate under the truck. Follow each model's instruction manual.

Some models are designed to have their circuit boards or wiring harnesses replaced by circuit-board style decoders, **6-3**. Although this might seem complicated at first, it really isn't. Installing a new board is simply a matter of removing the original board, setting the new board in place, and replacing the connectors. Decoder manufacturers generally list the appropriate locomotives compatible with their decoders.

Decoders are available from several manufacturers, including Atlas, Bachmann, Digitrax, ESU, Lenz, MRC, NCE, QSI, and Soundtraxx. Most manufacturers offer inexpensive basic decoders designed to just operate the motor and turn the headlights on and off. More advanced decoders have additional function outputs for advanced controls, or they can power additional features such as ditch lights, rooftop strobe lights, or warning lights.

Sound decoders, as the name implies, allow locomotive-related sounds to play through a speaker mounted in the model, **6-4**. For diesels, these usually include horn, bell, and engine sounds (with variation in engine rpms as the speed changes), as well as dynamic brake whine. Steam units supply whistle, generator, and steam blow-off sounds along with chuffs that vary with engine speed. Many sound decoders match specific prototype locomotives, are programmable with downloadable sounds, and include other train noises such as coupler crash.

Sound decoders are available as combination units that control both the motor and sound. They can also be sound-only units designed to be added to a model already equipped with a decoder or used with an unpowered (dummy) locomotive.

Installing a sound decoder follows the same process as with a standard decoder, but more room is needed for the speaker. Check the manufacturer's instructions for details and specific installation instructions.

Many high-end locomotive models come with decoders, **6-5**, and even sound units with speakers already installed, **6-6**. Most of these models are also designed to operate on standard DC power when not running on a DCC layout, and many of the sound features can be operated (with more limited controls) with a standard DC power pack.

Many newcomers to the hobby pass on DCC because of the initial cost or because they think it's only for large layouts. I encourage you to consider DCC because of the versatility in operations it provides and the ease in operating added features such as sound and lighting. And, if you eventually build a larger layout, you'll find the cost of a DCC system is offset by the expense of switches, wire, and other components needed for multiple-train wiring using standard DC power. Finally, a DCC system is far more flexible and easier to operate.

Basic maintenance

New locomotive models are about as maintenance-free as they can possibly be. About all that's necessary is to lubricate the wheel bearings and gears in each truck every six to 12 months or after several hours of operation.

To lubricate, first remove the cover on the bottom of the truck, **6-7**. Be sure to check the instructions when doing this, as manufacturers use many methods for securing the cover.

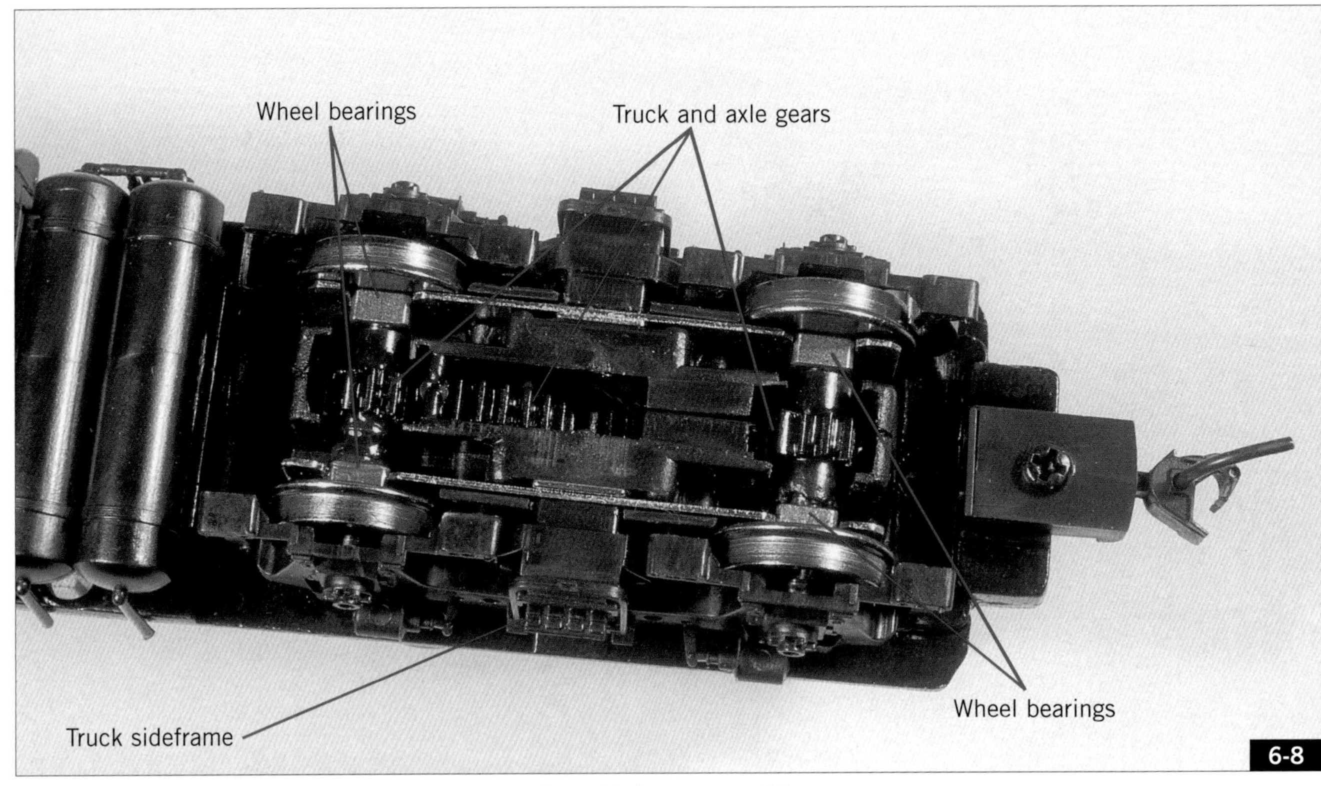

Add a drop of light oil to the wheel bearings and a drop of gear lubricant to one of the gears.

For bearings, use a light oil such as Atlas ConductaLube, LaBelle 108, or Woodland Scenics Ultra-Lite Oil; for gears, use a heavier lubricant such as Atlas no. 190, LaBelle no. 102, or Woodland Scenics Gear Lube, **6-8**. A single drop on one gear is enough; the movement of the gears will carry the lubricant throughout the gearbox.

Many locomotives come from the factory with excessive lubrication in the truck areas – check each new model for this as you take it from the box. Excess oil and grease will attract dust and dirt, hampering operation, so remove it with a cloth or paper towel.

Clean wheels are important for electrical contact. You can clean wheels by placing a paper towel across the tracks and wetting the areas on each rail with track cleaner, **6-9**. Clean one truck's wheels at a time by applying power and holding the locomotive in place with one truck over the paper towel and the other on the rails. The wheels will clean themselves as they spin on the towel. Then repeat the process with the other truck.

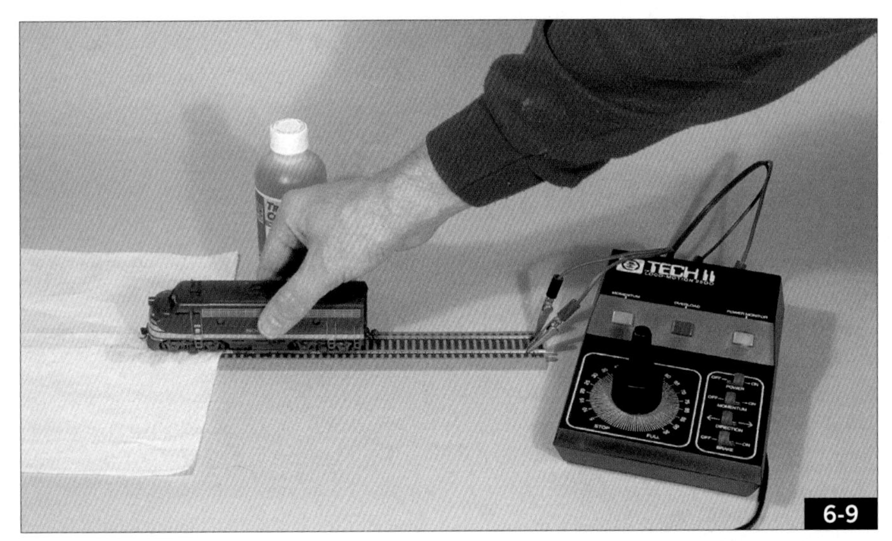

Run a locomotive on a paper towel wet with track cleaning fluid to clean its wheels. Do one truck at a time.

Rolling stock

Freight and passenger cars have enjoyed the same evolution in quality as locomotives. Into the 1990s, most ready-to-run rolling stock was aimed at the toy train or train-set markets. Cars weren't highly detailed and had many details molded in place. Any separately applied detail items were heavy and out of scale. Kits were the way to get detailed, realistic cars.

These days, realistic, high-quality ready-to-run cars are offered by many manufacturers, including Accurail, Athearn, Atlas, Bowser, Branchline, InterMountain, Kadee, Proto 2000, Red Caboose, and Walthers. Most of these cars feature separately applied details including grab irons, ladders, etched running boards, and brake gear. Along with realistic details, these cars run well with smooth-rolling metal wheels and knuckle couplers, **6-10**.

With the 18"-radius curves on this layout, 40- and 50-foot freight cars are the best choice for rolling stock.

6-10

Kadee's PS-1 boxcar was the first high-end ready-to-run freight car. It features many separate details with fine molding, metal wheels, and knuckle couplers.

6-11

A Kadee coupler height gauge is handy for checking that your couplers are all installed properly. Low couplers often uncouple accidentally, and low-hanging pins can snag at turnouts and crossings.

Good ready-to-run passenger cars have also become widely available. Many passenger car models now include interior detailing, working end diaphragms, and underbody detail.

Couplers and wheels

Automatic knuckle couplers are now standard on almost all new equipment, **6-11**. These are realistic, couple together easily, and can be automatically uncoupled using magnetic uncoupling ramps, **6-12**, or hand uncoupling tools, **6-13**.

The metal pin that extends from the bottom of the coupler is the uncoupling pin. When a car is placed over a magnetic uncoupling ramp, the magnet pulls the pins on adjoining cars to the sides, which opens the knuckles. Separation won't happen if the cars are pulled taut (for example, if the train is moving) but will occur if there's slack between the cars.

On a small layout like this one, where uncoupling moves are within easy reach, I recommend simply using a hand uncoupling device instead of magnets. Hand devices are placed between the coupler knuckles and twisted, which opens the knuckles and separates the cars.

Portable uncoupling magnets, such as the Kadee no. 321, work by pulling the uncoupling pins on adjoining cars to the sides. They can be picked up and placed wherever needed on a layout.

Make sure that the uncoupling pins on each car are tall enough to clear between-track items such as turnout closure rails and grade crossings. A Kadee coupler height gauge is the ideal testing tool. If a pin hangs low, it can be bent upward.

Most advanced modelers prefer metal wheels on rolling stock. Metal wheels are heavier than plastic wheels, so they lower a car's center of gravity and improve operation. Metal wheels also help polish and clean the track, and because of this, they generally stay clean. Plastic wheels tend to gather dust and gunk from the rails and become dirty over time. The best solution for this is to replace plastic wheelsets with metal replacements. These are made by Inter-Mountain, Kadee, NorthWest Short Line, ReBoxx, and others.

Accurail's Switchman and other uncoupling tools work by opening the knuckles between cars.

Realism and operation

Along with structures, locomotives and rolling stock help place your layout in an era and region. Many model railroaders have a favorite railroad, and they stick to locomotives and cabooses painted and lettered to match. Freight cars can be from railroads around the country, but usually, more "home road" cars should be present than "foreign-line" cars. For example, if you model the Chicago & North Western, about half the cars should be C&NW, 25 percent from surrounding railroads, and the rest from other lines around the country.

For best operation, in general, the tighter the curves on a model railroad, the smaller the equipment that should be used. Longer locomotives and rolling stock will sometimes derail because of the excessive end overhang on curves, **6-14**. And even if a longer car or locomotive will run, it often doesn't look realistic on tight curves.

Since this layout is set in the 1960s, I kept the equipment in that era, with Athearn Genesis F units, a Proto 2000 Alco RS-27, and a Proto 1000 Alco RS-2, along with a Proto Heritage 0-6-0 steam switcher. The freight cars are mostly 40-footers from several manufacturers. The combination results in reliable operation and a more realistic appearance, **6-15**.

In the last chapter, let's look at adding some final details and finish the layout – if one can ever truly be called finished.

6-14

Long cars, such as the scale 89-footer on the left, don't operate well on tight curves. The excessive end overhang can lead to derailments.

6-15

A four-axle Chicago & North Western Alco RS-27 moves a pair of hopper cars (from Accurail and Athearn) toward the coal trestle.

7-1

CHAPTER SEVEN

Final details

There's a wealth of opportunities for detailing even the smallest layouts, including signs, vehicles, figures, fences, and other items.

Although our layout is essentially done, many final details, such as vehicles, people, and signs, need to be added yet, 7-1. Adding final details is easy and fun to do. Along with structures, locomotives, and rolling stock, these detail items help set the time and location of a layout and give it character – and characters! You can create various mini-scenes, and if you get bored with one, you can just swap out details to create a new one. In fact, it is this possibility of continuous detailing that keeps many modelers involved in the hobby, and it's what keeps a layout from stagnating or ever being completely finished.

Loading dock details from Kibri, barrels from JL Innovative, vehicles from Mini-Metals, and a Woodland Scenics figure complete this scene.

Just a few of the nicely detailed vehicles on the market include this firetruck from Athearn and autos from Mini Metals and Woodland Scenics.

Several vehicles wait at a gate crossing as a train passes.

Details, details

A quick trip through a hobby shop or a look through the Walthers catalog will reveal a tremendous variety of detailing possibilities, including vehicles, figures, signs, streetlights, fire hydrants, trash containers, barrels, soda machines, crates, and other items, **7-2**. Many of these items are available ready to use, and most are very straightforward to install. See what's available, use your imagination, and detail your layout to match your interests.

Vehicles and figures

High-quality assembled North American vehicles are made by Athearn, Atlas, Boley, Busch, Classic Metal Works (Mini-Metals), Con-Cor, First Gear, Fresh Cherries, Herpa, IHC, MotorArt, Norscott, Ricko, Trident, Woodland Scenics, and others, **7-3**. The selection includes automobiles, on- and off-road trucks, construction equipment, and military vehicles. Most of these vehicles are based on actual trucks and cars from the 1930s to the present.

Putting vehicles on a layout is usually a matter of placing them on roads and parking areas where you want them. There's usually no reason to glue

A tiny dab of putty under a wheel holds a vehicle in place on a hill.

7-5

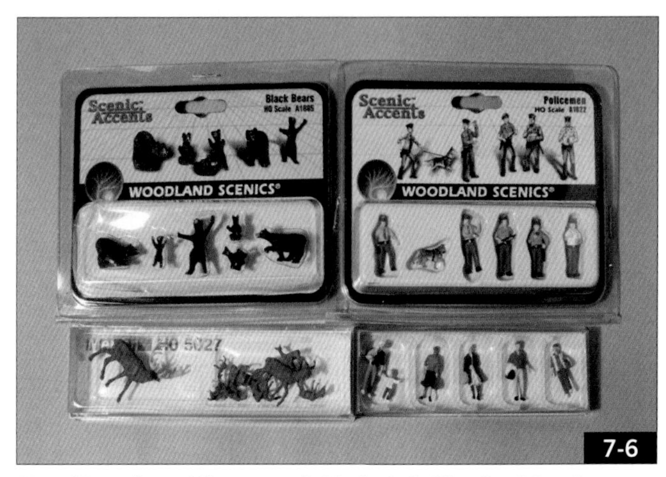

The wide variety of figures available include Woodland Scenics bears and policemen with a dog, Merten deer, and Preiser walking travelers.

7-6

Figures help bring life to a layout. Be sure to place them in logical places and groupings.

7-7

Touch the feet of a figure lightly in a drop of cyanoacrylate adhesive (CA).

7-8

Place the figure and hold it a few seconds until the glue sets.

7-9

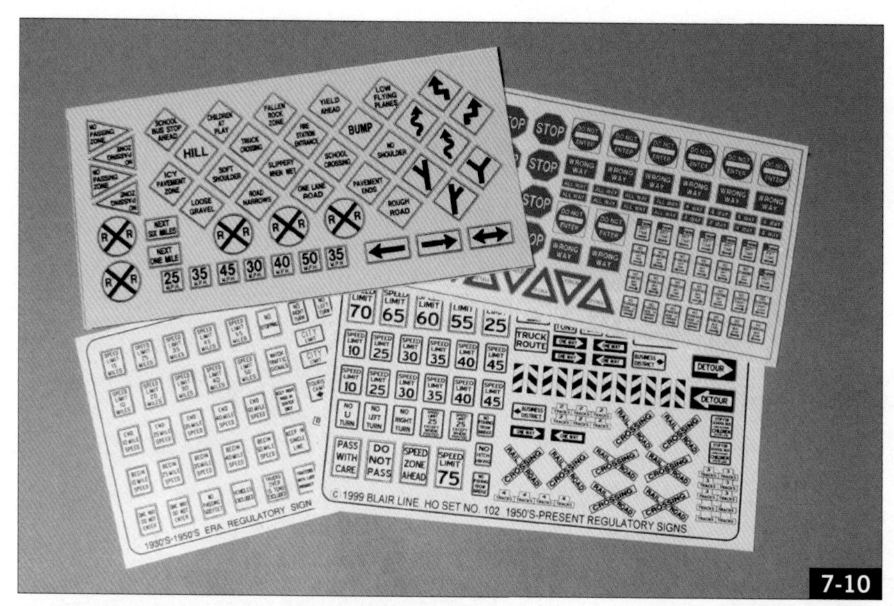

7-10

Sets from Blair Line include yellow and red highway warning and regulatory signs as well as steam-era and modern speed limit and warning signs.

7-11

Poke a hole in the scenery at an appropriate spot and press the signpost into place.

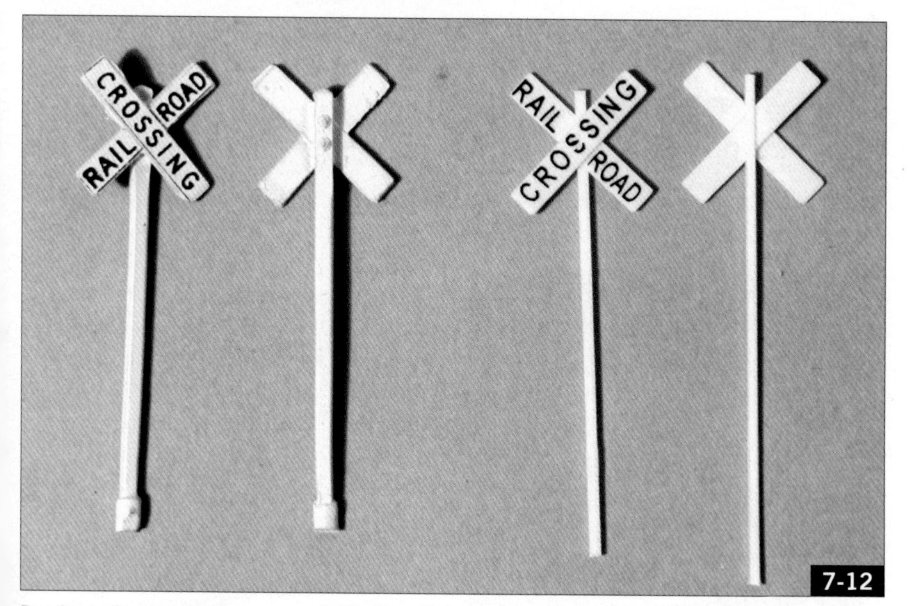

7-12

Ready-made crossbucks are available from JL Innovative Design (left) and CMA (right).

or otherwise secure them in place – this makes them easy to move around for photos or to give the layout a different look, **7-4**.

If you place a vehicle on a hill, a small dab of putty under a tire will hold it in place, **7-5**. Just keep the putty out of sight on the side of the vehicle away from viewing lines.

Adding miniature people to the layout makes a model railroad come alive. Painted figures are available from Faller, Merten, Preiser, Woodland Scenics, and other manufacturers, **7-6**. You can find almost any type of figure needed, including pedestrians, seated people, railroad workers, wedding attendees, photographers, gas station attendants, construction workers, passengers, cyclists, sports players, diners, and drivers. They are offered in a variety of poses and clothing styles to match eras from the 1800s through today, **7-7**.

You can also find a wide selection of animals: pets such as dogs and cats; farm animals including cattle, pigs, and sheep; and wild animals like bears, moose, and deer.

To permanently plant a figure, dip its feet lightly in a drop of medium-viscosity cyanoacrylate adhesive (CA), **7-8**. Touch the feet to a scrap piece of plastic to remove most of the glue and set the figure carefully in its intended location. You can use your fingers, or if clearance is tight, tweezers. Hold the figure for a few seconds until the glue sets, **7-9**. For temporary placement of figures, you can use rubber cement.

Make sure figures are placed in realistic positions and locations. Figures that show motion are fine in photos, but when viewing a layout, static figures, such as two people having a conversation on a sidewalk or a child peering into a storefront window, look better.

Signs

Highway and railroad signs offer many opportunities for detailing a model railroad. Look carefully the next time you're on the highway or walking down a street, and you'll become aware of hundreds of signs, including speed limit, route, junction, and warning signs, as well as billboards and other advertising signs and business signs. Rail-side signs include whistle posts, mile markers, speed signs, and many others. Many signs are ready to use, but some may need a little assembly and finishing.

Adding highway signs to a layout is simple, with a wide variety available from Blair Line, **7-10**, and other

The billboard is a JL Innovative sign on a Walthers frame. The speed sign is from Blair Line, and the crossbucks are from CMA.

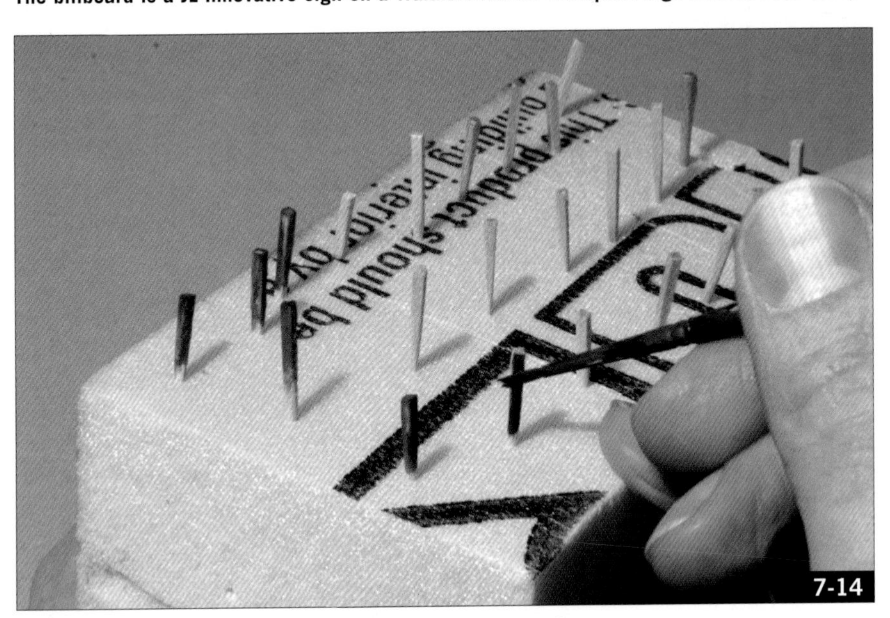

Paint the toothpick fence posts with thinned grimy black paint.

manufacturers. The Blair sets include square stripwood pieces for posts. Paint these with a wash of Polly Scale grimy black paint (about one part paint and seven or eight parts water or Polly S Airbrush Thinner). Cut the sign from the sheet with a sharp hobby knife and use CA to glue it to the post. Use an awl to poke a hole into the scenery at an appropriate location and push the sign in place, **7-11**.

Crossbucks – the X-shaped railroad crossing signs found at every public grade crossing in the country – are key detail items. They have appeared in many styles over the years and vary by railroad and region. Modern crossbucks are set at a 90-degree angle, but earlier signs often had the panels crossing at a shallower angle. Several companies offer these signs in ready-made versions, including JL Innovative Design and Creative Model Associates (injection-molded styrene), **7-12**. CMA also makes injection-molded, ready-to-install circular orange crossing warning signs as well. If you're looking for crossbucks with warning lights and gates, they're offered by NJ International, Tomar, and Walthers.

Billboards are another attention-getting type of sign, displaying a wide variety of advertising subjects such as autos, food products, tobacco, beer, railroads, oil, and other businesses. Made by Blair, JL Innovative, and others, many are available in kit form, but Walthers offers an assembled frame (no. 933-3133) with blank sign that can be used with any manufacturer's sign. I added a JL Innovative automobile billboard to one of these frames, cutting it to size and applying it with double-sided mounting tape, **7-13**.

7-15

Push the posts into place in the scenery and secure them with a drop of glue.

7-16

Glue fence line to the posts with a bit of cyanoacrylate adhesive (CA).

7-17

ome assorted detail items, from left, include barrels, soda cases, and an oil rack from JL Innovative; an oil rack from Walthers; a soda achine kit from Woodland; and a Coke machine from Athearn.

As with structure signs (chapter 3), remember that any sign you can photograph you can turn into a model – billboard, highway sign, regulatory sign, or poster.

Miscellaneous details

To complement the plowed field on the left side of the layout, I decided to add a cow pasture to the right side. This required a fence to hold in my Preiser cow figures. You can easily make a low-cost farm fence in a few minutes with square toothpicks. Use a hobby knife to cut each toothpick in half and stick them into a piece of foam, **7-14**. Paint them medium to dark gray with a wash of grimy black paint (similar to the sign posts). I tried to vary the intensity and appearance of the posts, giving some a dark coat and some a light wash.

Dip the ends in a bit of white glue and push the posts into the scenery as needed (you may have to poke a pilot hole with an awl), so they stick out of the ground a scale four feet or so, **7-15**. I made the fence wire from flexible elastic thread called E-Z Line made by Berkshire Junction, but common thread will work as well. I used a rust color, but you can also use black or gray. String the line from post to post, placing a dab of CA on each post and holding it in place with a tweezers, **7-16**.

Complete the scene with some cows or steers (or the farm animal of your

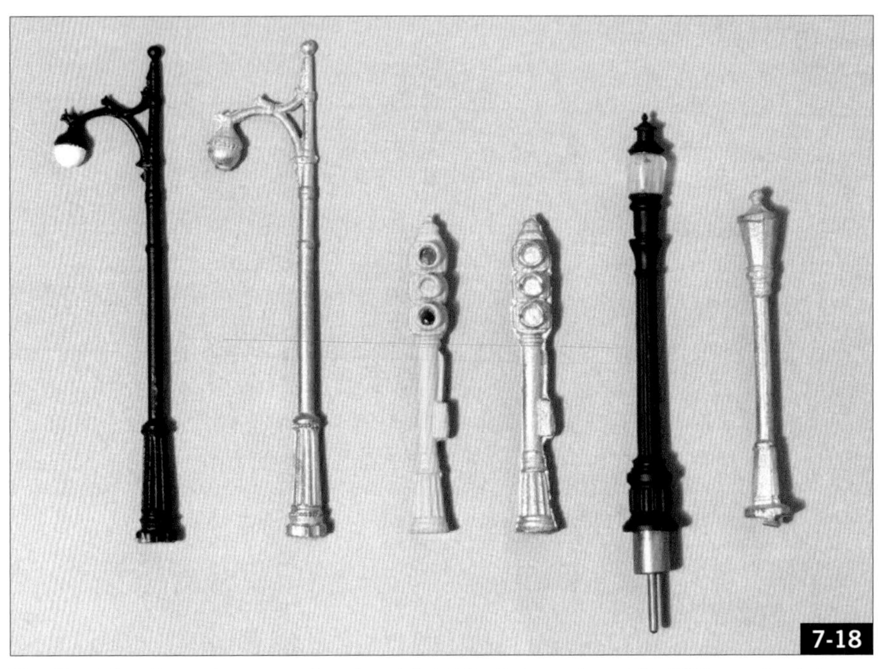

7-18

Woodland Scenics makes several cast-metal details that require a bit of filing and paint, including the streetlights and stoplight shown here. Walthers offers illuminated streetlights (second from right) in several styles.

7-19

Pikestuff's highway guardrail is an example of an easy-to-assemble detail item.

7-20

Mark mounting locations for the guardrail posts, make the holes, and push the guardrail into place.

choice). My holsteins are from Preiser (set no. 14155). Add a small drop of CA on each hoof and press the animal into place.

Thousands of small detail items are available to dress up structures, loading docks, sidewalks, station platforms, and other areas, **7-17**. Several of these are ready to add or install right from the box; others my require a bit of filing, if not cast cleanly, and some paint, **7-18**. I used some nonoperating Woodland Scenics streetlights on the right side of the layout, and three powered Walthers streetlights on the left side (although for now they're not wired).

Other details come as simple kits. Don't be afraid to tackle these – usually a bit of glue, some paint, and a little time will produce a nice item for the layout. The Pikestuff highway guardrail kit is an example, **7-19**. I thought the steep embankment on the right side of the layout warranted some protection along the highway. Assembling the kit was simply a matter of gluing the railing to the posts. I then painted the whole thing light gray. To install it, set the assembly at the proper location and make mounting holes with a scriber or an awl, **7-20**. Then, press the guardrail into place and you're done.

Walthers track bumpers are another good example of simple detail kits, **7-21**. Every spur track needs a bumper of some type to keep cars from rolling off the ends. These simple plastic kits consist of three parts that must be glued together. After a bit of dark brown paint, they're ready to glue in place between the rails, **7-22**.

After you add the last of your final details, step back and look over your layout. You'll notice areas where you solved problems and scenes that look great. And you may notice that there are things that you would like to modify or do over. And the great thing is that you can, **7-23**. That's the fun of model railroading.

Use your imagination as you seek out ways to improve, detail, and upgrade your layout, and you'll find it to be true that no model railroad is ever truly finished.

Walthers track bumpers are easy-to-assemble plastic kits.

Glue a bumper between the rails at the end of each spur track.

Even though this layout is now complete and ready to run, you can always add or change details, add structures, or find other improvements to make.

Resources

The following companies are model railroading suppliers featured in the book. They are listed with a general product category and its applicable chapter. Many of them offer additional products, which are described in more detail on their web sites.

Accurail
Elburn, IL
www.accurail.com
Rolling stock, chapter 6

AIM Products
Schofield, WI
www.aimprodx.com
Scenery, chapter 4

American Art Clay Co.
Indianapolis, IN
www.amaco.com
Sculptamold, chapter 5

American Model Builders
St. Louis, MO
www.laserkit.com
Structure kits, chapter 3

Athearn
Carson, CA
www.athearn.com
Locomotives and rolling stock, chapter 6

Atlas Model Railroad Co.
Hillside, NJ
www.atlasrr.com
Track, chapter 2; Locomotives and rolling stock, chapter 6

Bachmann Industries
Philadelphia, PA
www.bachmanntrains.com
Locomotives and rolling stock, chapter 6; Scenery, chaper 5

Blair Line
Carthage, MO
www.blairline.com
Signs and billboards, chapter 7

Bowser
Montoursville, PA
www.bowser-trains.com
Locomotives and rolling stock, chapter 6

Central Valley
Oceano, CA
www.cvmw.com
Structures, chapter 3; Scenery, chapter 4

Chooch Enterprises
Maple Valley, WA
www.choochenterprises.com
Scenery, chapter 4; Details, chapter 7

City Classics
Pittsburgh, PA
cityclassics.fwc-host.com
Structure kits, chapter 3

Classic Metal Works
Sylvania, OH
www.classicmetalworks.com
Vehicles, chapter 7

Con-Cor
Tucson, AZ
con-cor.com
Locomotives and rolling stock, chapter 6

CVP Products
Richardson, TX
www.cvpusa.com
DCC systems, chapter 6

Design Preservation Models
Linn Creek, MO
www.dpmkits.com
Structure kits, chapter 3

Digitrax
Panama City, FL
www.digitrax.com
DCC systems, chapter 6

Heki (see Wm. K. Walthers)
Details, chapter 7

Hot Wire Foam Factory
Lompac, CA
www.hotwirefoamfactory.com
Foam-cutting tools, chapter 4

InterMountain Railway Co.
Longmont, CO
www.intermountain-railway.com
Rolling stock, chapter 6

JL Innovative Design
Sauk Rapids, MN
www.jlinnovative.com
Signs and details, chapter 7; Structure kits, chapter 3

Kadee
White City, OR
www.kadee.com
Rolling stock, chapter 6

Kato USA
Schaumburg, IL
www.katousa.com
Locomotives and rolling stock, chapter 6

Micro-Mark
Berkeley Heights, NJ
www.micromark.com
Tools, chapter 2

Model Rectifier Corp.
Edison, NJ
www.modelrectifier.com
Train control, chapter 6

Proto 2000 (see Walthers)
Locomotives and rolling stock, chapter 6

Red Caboose
Mead, CO
www.red-caboose.com
Rolling stock, chapter 6

Rix Products
Evansville, IN
www.rixproducts.com
Structures, chapter 3; Details, chapter 7

Scenic Express
Jeannette, PA
www.scenicexpress.com
Scenery, chapter 5; Details, chapter 7

SoundTraxx
Durango, CO
www.soundtraxx.com
DCC decoders, chapter 6

Testor Corp.
Rockford, IL
www.testors.com
Paints, including Polly Scale, chapters 3 and 4

Wm. K. Walthers
Milwaukee, WI
www.walthers.com
Structures, chapter 3; Rolling stock, chapter 6; Details, chapter 7

Woodland Scenics
Linn Creek, MO
www.woodlandscenics.com
Structures, chapter 3; Scenery, chapter 5; Details, chapter 7

About the author

Jeff Wilson has written numerous books on railroads and model railroading. Jeff is a freelance writer, editor, and photographer who contributes to various magazines including *Model Railroader,* where he spent 10 years as an associate editor. He also writes a model railroading column for *Model Retailer* magazine and is a correspondent for *Trains* magazine. He enjoys many facets of the hobby, especially building structures and detailing locomotives. He also enjoys photographing both real and model railroads.

Acknowledgments

Several manufacturers were kind enough to supply materials used in this project. Thank you to Bob Walker of Accurail, Mark Ballschmieder of AIM Products, Tim Geddes of Athearn, Laura Kolnoski of Atlas, Dale Rush of Blair Line, Jack Parker of Central Valley, Jim Sacco of City Classics, Mike O'Connell of Chooch, Zana Ireland of Digitrax, Dave Proell of JL Innovative Design, Tom Piccirillo of Micro-Mark, Rick Rideout of Rix Products, Sermeng Tay-Konkol of Walthers, and Cody Gilmore of Woodland Scenics.

Add realism to your layouts

Learn how to improve the appearance of your diesel locomotives and thus the realism of your layouts. You'll learn how to add essential details like horns, hoses, and grab irons and how to open engine-room doors and add cab detail, upgrade couplers and wheels, and bring it all together with decals and paint.

12421 • $19.95

Improve the appearance and performance of your freight cars! Learn how to improve ready-to-run and kit HO and N freight cars with upgraded details, wheels, couplers, and loads, and how the expert tips apply to your own HO, N, O, and G scale cars.

12420 • $18.95

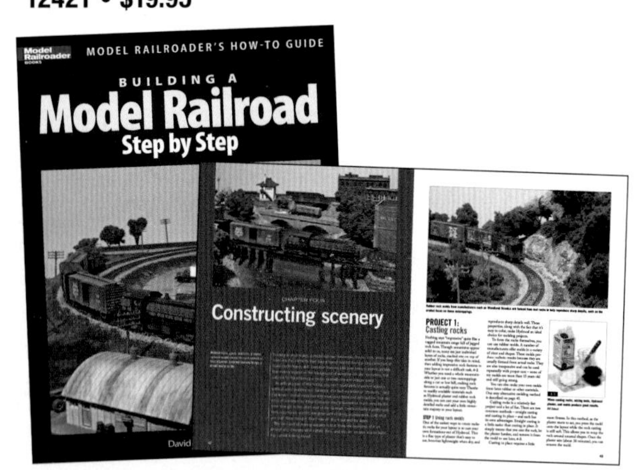

Model Railroader magazine managing editor David Popp takes you through the complete construction story of his N scale Naugatuck River Valley layout with more than 30 step-by-step projects detailing every phase of construction. Also covered are choosing a scale, working in your available space, and installing DCC.

12418 • $21.95

Learn how easy it can be to add a lifelike backdrop to your model railroads. Painting Backdrops for Your Model Railroad lets you know what you'll need to paint a backdrop, how to build and blend it into their layout scenery, and offers instructions on how to build specific backdrops with themes such as Midwestern, mountain, desert, city or town, and even how to add clouds. It's a must-have guide for model railroaders wanting a realistic background.

12425 • $18.95

ModelRailroaderBooks.com

Available at hobby shops.

To find a store near you visit www.HobbyRetailer.com

Call 1-800-533-6644

Monday–Friday, 8:30 a.m.–5:00 p.m. Central Time. Outside the U.S. and Canada, call 262-796-8776.

PMK-BKS-12429RH